Guide to Personal Knowledge

The Philosophy of Michael Polanyi:
Tacit Knowledge, Emergence and the Fiduciary Program

Dániel Paksi
Department of Philosophy and History of Science,
Faculty of Economics and Social Sciences,
Budapest University of Technology and Economics,
Műegyetem rkp. 3., H-1111, Budapest, Hungary

Mihály Héder
Department of Philosophy and History of Science,
Faculty of Economics and Social Sciences,
Budapest University of Technology and Economics,
Műegyetem rkp. 3., H-1111, Budapest, Hungary
ELKH SZTAKI

Series in Philosophy

VERNON PRESS

www.vernonpress.com

In the Americas:
Vernon Press
1000 N West Street, Suite 1200
Wilmington, Delaware, 19801
United States

In the rest of the world:
Vernon Press
C/Sancti Espiritu 17,
Malaga, 29006
Spain

Series in Philosophy

Library of Congress Control Number: 2021943082

ISBN: 978-1-64889-313-1

Table of contents

List of figures and tables

Foreword

C. P. Goodman

Epistemology is the study of knowledge claims. Some assert we ought to put two baskets in front of us. In the left basket, we should put logic, reason, and thought and call it the context of justification. Into the right basket, we should put history, psychology, and experience and call it the context of discovery. Polanyi rejects the assumption that the context of justification can be isolated from the context of discovery. While he endorses the epistemological turn which characterizes modern philosophy, he qualifies it by denying that we can isolate epistemological claims from metaphysical assumptions. The Logical Empiricists declared that any claim which cannot be derived from logic or experience has no epistemological value. But if this is correct, then this declaration has no epistemological value. Every intellectual journey starts from the first step. Philosophy investigates this phase. In the history of philosophy, efforts have been made to secure this first step. In "The Unity of Philosophical Experience" (1937), Etienne Gilson notes that various Western thinkers have sought to replace philosophy with knowledge derived from other disciplines, such as logic or mathematics, history or psychology, theology or physics. But philosophy always buries its undertakers. Every attempt to secure a foundation relies upon philosophical assumptions. Nor should these assumptions be ignored. As thinking beings, we ought to reflect upon the foundations of our beliefs. Michael Polanyi (1891-1976) was a Christian, a scientist, and a liberal. Many (but not all) of those who have written about his ideas have been Christians, scientists, and liberals, sometimes all three. But Polanyi was also a pantheist, a humanist, and a conservative. Maybe these claims do not hold together, and while he is one of the most important philosophers writing in English in the middle years of the C20th, his ideas are rarely studied as a whole, rather he is mentioned in footnotes as somebody who has made significant contributions to discussions about polycentric spontaneous order, the acquisition of science as a moral practice, and the shifts in understanding which take place in scientific change. It is possible to detect his influence in the writings of Hayek, Kuhn, and Feyerabend. Polanyi denies that science is a practice that can simply be reduced to rule-following, but above all, he rejects the concept of knowing without a knowing subject. Knowing is neither objective nor subjective; it is personal and participatory. His epistemological approach is neither empiricist nor idealist. All knowing is grounded in experience, but all experience is interpreted. We rely upon a body, but we also indwell within articulations. We strive to provide descriptions of what is real, but the descriptions themselves are a real experience. We rely upon more than we can say, and we say more than we can know. The assumption that there are two quite separate cultures, humanistic and scientific, is based on a false division between the subject and the object. All knowing is personal. Nor are we isolated; we exist within communities. Indwelling within articulations enables us to formulate transcendent ideals, which in turn generate communities sustained and constrained by their shared beliefs. We acquire traditions. This emphasis upon inherited practices, which can also be found in the later Ludwig Wittgenstein, influenced Charles Taylor and Alistair MacIntyre. But ideas do not exist independently of the person who thinks about them. Recognizing the power of ideas does not require us to deny the reality of the person. We are not, as John Lukacs points out in "Last Rites" (2009), simply vehicles for ideas; we make choices. Nor is Polanyi a cultural relativist. We are embodied. Every conscious being generates meanings by relying upon a hierarchy of boundary conditions.

As he lays bare the structure of our tacit awareness, Polanyi examines the phenomenology of tool use, the tacit ground of meaning, and the implications these insights have for computing. His writings influenced Hubert Dreyfus and inspired some of the engineers at Xerox Parc. We use symbols as vehicles for meaning, but for symbols to become meaningful, they have to be used by conscious embodied agents. Polanyi rejects the fact/value distinction, highlights the importance of moral passions, and repudiates the phenomena of moral inversion. Rendered homeless by a false conception of science, moral passions are diverted into utopian forms of nihilism. Instead of knowing viewed as the desire to reduce the number of untruths and justice viewed as the desire to reduce the amount of injustice, intellectual passions are viewed as expressions of nothing more than expressions of power. Polanyi rejects both idealism and materialism. His theological and teleological integration of transcendence and pantheism is a corollary of his rejection of subjectivism and objectivism. God is disclosed in emergence and is both created and discovered. Critical Philosophy is the belief that enlightened people do not pursue God; they seek to direct their passions in accordance with correct methods. This is a lineal descent of the claim that we can secure a ground for our beliefs. While denying that we can reach an absolute perspective, Critical Philosophy assumes that we can know the apparatus through which knowing becomes possible. This approach dominated Western philosophy between Locke and Kant, and despite its deconstruction by Hegel, it lives on in Analytical Philosophy. It is via an appeal to scepticism that knowledge claims are designated as nothing more than assertions of power. This claim, however, is excluded from any sceptical denunciation. This hypocrisy should not shock us. It is not uncommon for egalitarians, for example, to declare their belief in egalitarianism while at the time assuming, on the grounds of their superior critical insight, their entitlement to a position of power. This entitlement is justified via an appeal to critical method, with other claims to privilege, such as appeals to birth or education, rejected. With revolutionary zeal, they seek to destroy every existing social and metaphysical order and replace it with a new society. In his "Plan of Scientific Studies Necessary for the Reorganization of Society" (1822), Auguste Comte (1798-1857) envisaged scientists benevolently directing society with a view to maximizing its happiness. A Post-Critical philosophy, however, takes a different approach. It points out that all knowledge is situated, but for exactly the same reason, all knowledge claims are fallible. Knowing is a process in which conscious beings rely upon and are guided by, and articulate and reflect upon, their tacit awareness, arriving at convictions they believe to be true while acknowledging that they may be false. Polanyi defends a free society not on the grounds that all claims are equally valid and that everybody is equally virtuous, but on the grounds that we do not have a philosopher's stone in our possession which enables us to infallibly distinguish between that which is true and that which is false. The good is an object of continual inquiry. Polanyi, in other words, supplies us with a heuristic philosophy. It is, I suggest, a philosophy for our time. Which is to say it is an account whose time has come. An increasing number of studies of his work are being written; this book is one such contribution. In this preface, I seek to briefly sketch out the background to some of his ideas.

Michael Polanyi was Hungarian. After a century and a half of Turkish rule, Hungary was left plundered and depopulated. A quarter of a century after the Turks left, the population of Buda, a cluster of modest houses and vineyards, was less than 13,000, and the population of Pest, which the historian John Lukacs describes as a semi-oriental fishing village, was little more than 4000. In the Eighteenth Century, the traders, financiers, and manufacturers, who lived in Buda were German-speaking, Catholic, and loyal to the Habsburgs. In the country, the Magyar-speaking gentry despised commerce, were often Calvinist, and opposed the Habsburgs. In the early years of the nineteenth century, in a mood of patriotic fervour, poets and scholars refined and enriched Hungarian into a great literary language. The Prussian defeat of the Austrians in 1866 meant that in 1867 most of the changes which Hungarian protesters had agitated for in 1848 were granted, although foreign affairs, defence, and economic policy continued to be directed from Vienna. In the last third of the nineteenth century, Budapest was the fastest-growing city in Europe. Its financiers, traders, and manufacturers were now often Jews, who Magyarised their surname, and were encouraged to emigrate from such places as Galicia in order to expand the Hungarian-speaking population. As a consequence of reforms begun under the Austrians, but continued by Hungarian ministers, schools in Budapest, following German and French models, began to set high standards of academic excellence. There was a fascination with science. Polanyi published his first scientific paper while he was still attending the Minta, his high school gymnasium. There was also much interest in how discovery takes place in science. Plato, in his "Meno" dialogue, and Saint Augustine, with his concept of "faith seeking understanding", both sought to understand how knowing is possible. As Polanyi put it, either know what you are looking for, in which case you already know it, or you don't know what you are looking for, in which case how can you find it? Although the Greek mathematicians Pappus and Proclus in Late Antiquity, and the philosophers Descartes and Leibnitz in the Early Modern period, had written about the discovery process in mathematics, it was William Whewell (1793-1866) who coined the word heuristic. He derived it from the Greek heuriskein "to find", the first person singular perfect form of which is heureka "I have found it". In his "On the Philosophy of Discovery" (1860), Whewell sought to steer a middle path between Bacon and Leibniz, claiming that facts and theories should be characterized in terms of opposition rather than difference. They are poles that change as theories are transformed into facts. Whewell denied that science has any special method. Knowing is like learning to walk; you take a few steps, stumble, and try once again to stride forward. Charles Pierce (1839-1914), taking Whewell's side against Mill, coined the term "abduction" in 1867 to denote a non-deductive inference that is not inductive. In Austro-Hungary, Bernard Bolzano (1781-1848), in his "Theory of Science" (1837), wrote a long chapter about the art of discovery in science. In Hungary, Sandor Mikola (1871-1945) introduced heuristic thinking into the school curriculum. Imre Lakatos (1922-74) translated "How to Solve It" (1945) by George Polya (1887-85) (a school friend of Polanyi) into Hungarian. Its Hungarian title was "School for Thinking". Lakatos, who was also influenced by Hegel and Duhem, saw Polya as the father of heuristics in mathematical discovery. Because Karl Popper (1902-94) viewed Polanyi as an intellectual enemy, Lakatos corresponded with him in secret.

Michael Polanyi was Jewish. Since Jews could no longer worship God in his temple, they revered him through the study of holy scripture. These texts stressed the importance of ethical behaviour. In secular Jews, this often manifested itself as reverence for learning and a belief in Leftist politics. The 1867 political settlement was not only a national liberation for Hungary; it was also a liberation for the Jews. The economic and professional flourishing of some Jews within a liberal society, however, began to be resented. While nationalism unified Germany and Italy, it divided Austro-Hungary. In Great Britain, support for Hungary began to fade as liberal writers such as R.W. Seton-Watson drew attention to discrimination against non-Magyar minorities. Some Jews began to question the patriotism which had animated earlier reformers. The Catholic Church was often anti-Semitic. Many secular Jews responded by becoming enthusiastic defenders of the Enlightenment and the French Revolution. In 1908, his older brother Karl Polanyi (1886-1964) was the first president of the Galileo Circle, a student society that championed Social Science as a source of political reform. Radicals began to view the Anglophile liberalism of the previous generation as old-fashioned. It was asserted that Jews became financiers, traders, and manufacturers, simply because money-making was the only way they could gain respect in a world that was prejudiced against them. Inspired by Karl Marx (1818-83), some of the children of the newly flourishing Jewish middle classes speculated about the possibility of creating societies in which the inequalities generated by free markets would be replaced by centrally directed economies. Some children of wealthy Jews attacked their parents' prosperity, some children of poor Jews attacked their parents' Jewishness, and for many, rejecting the pursuit of wealth and rejecting Jewishness was the same thing. Marx looked forward to a day of violent judgement, at which the chosen people, the proletariat, will be liberated from the inequities of the market system. Expelling divine and ethical purposes from the world, Marx sought to replace them with economic determinism. Instead of the free markets, the new rulers, the enlightened, would decide who would get rewarded. Instead of a society based upon private property, autonomy would be replaced by coercion. A vision in which community takes priority, and a vision in which individual autonomy takes priority, are incompatible. Radicals and conservatives united in their opposition to liberalism. The Hungarian Jew Max Nordau, in the book "Degeneration" (1892), described the late Nineteenth century as a period of restlessness and fear of the future. Inspired by the Darwinian theory of evolution, Nietzsche declared that God is dead, and we should return to a morality based on survival of the fittest. Progressives were keen to express their contempt for the materialism of bourgeois life. Many greeted the prospect of war with enthusiasm. They saw it as an opportunity to restore comradeship and authenticity. As a reaction to the Dreyfus Affair, a Hungarian Jew called Theodor Herzel began to promote the view that there ought to be a Jewish homeland. Liberalism, he concluded, had failed, and he sought to free Jews (secular and otherwise) from the condition of being outsiders in a society which resented them. The Austrian novelist Herman Broch (1886-1951), however, wrote that it is reverence for life that is the supreme Jewish virtue. Jews torment themselves by conceiving God as transcendent, but it is incumbent upon them to embrace the pursuit of goodness, even in the absence of any hope of attaining it.

Bernard Bolzano (1781-1848) not only wrote about heuristics, he also wrote about language. He claimed that propositions exist prior to language. That they should be distinguished from the subjective acts which conceive them. The Austrian literary critic Fritz Mauthner (1849-1923) said we should purify language of its metaphysical assumptions. The reality of our experience can only be lived from moment to moment; it cannot be embalmed in words.

Influenced by Meister Eckhart (1260-1328), he substituted the concept of Tao for God. The Austrian satirist Karl Kraus (1874-1936) responded to Mauther, however, by asserting the reality of descriptions. All thought is language. Mauther declared that his interest in language was inspired by polyglot Austro-Hungary. The Hungarians, for a different reason, also took language very seriously. The Magyar language embodied the Hungarian spirit. Hungarians revered their poets even more than their scientists. Hungarian poets were influenced by the French Symbolist poets Baudelaire and Rimbaud. But the conception of art for the sake of art was not amenable to Hungarians, from whom literature historically was an extension of politics. In 1906 the Hungarian poet Endre Ady (1877-1919) published a book, "New Poems", which became a source of inspiration for those who believed that the process of political change begun by Nineteenth-Century reformers ought to continue. In 1908 a new literary journal called Nyugat (West) was launched, edited by the writer and literary critic Ernő Osvát, the aesthete Paul Ignotus, and wealthy patron of the arts Baron Sándor Hatvany-Deutsch, all of whom were Jewish. It defended a vision in which pursuit of the Good reconciles the ego with society and the universe. Atilla József (1905-37) captures the essence of this thought in his poem "Two Hexameters" (1936) "Why should I be good? They'll stretch out my corpse anyway! Why should I not be good! They'll stretch out my corpse anyway". In a poem commissioned by the materialist and atheist Galileo Circle, Ady declared that, life lives for sacred reasons, and if you are Hungarian, you have to strive a hundred times harder! Everybody ought to be given the opportunity to live their life to its fullest. All polarities are reconciled once we realise that God is everything. This was explained by his friend Hatvany in the following terms. Just as it is possible to reach down to the centre of the Earth from every point, so Ady seeks to reach down through every capricious idea, to the centre, God. In Ady, you find not only his system but also the opposing system because he believed that even the most encompassing system is too narrow for the universe. Ady insists on the personal character of experience and rejects conventional forms of worship "Show yourself to us, you are not Christian, you are not Jewish. You are the frightful terrible Lord." In Jewish mysticism, the potential of language to express thought, and the capacity of thought to comprehend what exists, are foundational. The principal preoccupation of Imre Madach's drama "The Tragedy of Man" (1860) is the question of whether the development of human thinking, guided by our pursuit of ideals, leads to happiness. His character Adam concludes that it does not, but he affirms life anyway. The struggle is the goal.

It was the German physicist Johan Lambert (1728-77) who introduced the term phenomenology to designate the study of appearances as distinct from the study of descriptions. In his "Psychology from an Empirical Standpoint" (1874), Franz Brentano (1838-1917) set himself the task of describing mental acts. He divides mental phenomena into three classes. First, presentations (vorstellung) occur whenever anything is present to consciousness. He revived the medieval term intentionality to denote the act whereby a mind intends its object. Second, judgement asserts or denies the existence of the object presented. Third, what is presented is attributed a positive or negative value. This determination is not subjective. He repudiates Bolzano's doctrine of presupposition-in-themselves, but he agrees with Bolzano that ethics is neither deontological nor utilitarian. The Good arises when we contemplate the whole, and we love it regardless of its attainability. His student Christian von Ehrenfels (1859-1932) examined the intentionality of identifying a whole, for example, a melody, that is more important than the parts within which it inheres. This was developed by Max Wertheimer (1880-1943) into a Gestalt theory of perception. Max Scheler (1874-1928) comprehends phenomenology as an ethical philosophy. In "Formalism in Ethics and Non-Formal Ethics of Values" (1913-16), he notes that Kant confines personhood to rational will. Scheler agrees with Pascal that objective values can be experienced emotionally by our intuition. The most valuable communities are those that rely upon empathy, but they also provide conditions that enable the most valuable persons to contribute. The chief threat to this is ressentiment, which, contrary to what Nietzsche asserts, is not Christian in origin but arises within societies that prioritise equality. In his last years, Scheler turned against Christianity on the grounds that it lifts man out of the cosmos. Defining God as the all-embracing whole, he turned instead to Eastern religions. In an outline "The Human Place in the Cosmos" (1928) for a book he did not live to complete, Scheler claims that God-in-becoming is realised through us because man is "world-open". We are not trapped in our environment like a snail in a shell. Nicolai Hartmann (1882-1950) agrees with Scheler that persons are necessary agents for realising values, but contrary to Nietzsche, he denies that we who create values, they enter our consciousness through axiological intuitions and exist in the deed. It is the "firmament of values" which are the primary agent; the free-acting person is a point on their journey from unreality to reality. Indeed it is only through the intrusion of values as determining powers that subjects morally become persons. Unlike Scheler, however, he claims that religions sublimate values in a way that abdicates our autonomy. Aurel Kolnai (1900-73) endorses the phenomenological view of ethics pioneered by Brentano and Scheler. According to Kolnai, communism is evil because it negates the self-directed individual personality. When Polanyi emigrated to England (after being appointed a chemistry professor at Manchester University), he countered the Marxist demand that science ought to be centrally directed with a Post-Critical epistemology in which cultural expression ought to be allowed to develop autonomously in accordance with their own laws. His focus of interest shifted from science to philosophy. In this preface, I have sought to supply some of the background to this new understanding.

Preface

This book aims to help the reader understand Michael Polanyi's most important book, titled *Personal Knowledge*, and as such to grasp the essence of his philosophical thinking. In this preface, Polanyi's goals are reconstructed first, and then his main philosophical arguments are introduced. The discussion is limited to the most crucial ideas, which are indispensable for the arc of his book. Given that Polanyi never restricted himself to one discipline, and frequently his comments draw on several areas, there is no doubt that alternate lists of important ideas could be created, drawing from specific perspectives such as those of epistemology, social theory, biology, etc.

The thirteen chapters of this book explain the essence of the thirteen chapters of *Personal Knowledge*. At the beginning of each of our chapters, we summarize Polanyi's goals and then provide a summary of the arguments. We include text boxes that provide background and other helpful information about scholars, concepts, or historic events cited by Polanyi. Also, we quote key passages from every chapter. Our book ends with an index and a list of quotations.

Acknowledgement

The writing of this book was supported by the János Bolyai Research Scholarship of the Hungarian Academy of Sciences. The thank the permission from the University of Chicago Press to quote Polanyi's *Personal Knowledge* in great length.

Polanyi's goal in writing *Personal Knowledge*

It is far from an overstatement to claim that Michael Polanyi's *Personal Knowledge: Towards a Post-Critical Philosophy* presents a unique worldview. Among other things, he touches on natural science, the arts and culture, evolution, social theory, history, engineering, mathematics, and religion. His expansive aim to cover almost the entirety of the human endeavor was a thing of the past even in 1958 when the book was first published, and his book did not resemble the more focused works common in the twentieth century. From Polanyi's perspective, however, work touching on such a diversity of disciplines was a consistent effort because, in his view, not only scientists but all modern people have a controversial and deeply troubling relationship with their own surroundings, goals, and ideals. The crisis of science, Polanyi's starting point, is not limited to science itself, since the origins of the scientific methods and the ideals of scientific discovery are not contained within science but are commonly present everywhere.

The phenomenon in Polanyi's focus is rooted in the distant past, and therefore it is important to give a historic dimension to the argument. Polanyi provides a description of the efforts of Western civilization to achieve knowledge of final and indisputable certainty, beginning in antiquity and continuing until the present in an array of changing manifestations. The most important of these changes, of course, happened in the early modern period, when modern science as the champion of objective investigations and certainty is created. The goal of modern science is to eliminate every personal element from knowledge; such personal elements are deemed subjective and thus should be discarded. This modernist epistemology provided extraordinary results if we compare the state of affairs to the centuries of the pre-Enlightenment era. And yet, a perfectly impersonal knowledge is a nonsensical ideal, in Polanyi's view. He believes that this development of the impersonal ideal of knowledge is the reason that modern human beings are so uniquely morally sensitive and are frequently morally outraged; modern moral passion is combined with excessive modern skepticism, and this undermines trust and traditional ideals such as truth and is responsible for the totalitarianism of the twentieth century. Moreover, the modern, free Western person is unprotected, both emotionally and intellectually, against totalitarianism, which after the Second World War means mainly Marxist-communist systems.

Polanyi's goal in writing *Personal Knowledge* was precisely to answer this twentieth-century trap by providing a concept of knowledge that enables modern persons to develop acceptable forms of relationship with older inherited traditions, and, at the level of the individual, self-acceptance, which includes a harmonic relationship to our human possibilities. Polanyi aims to help modern persons become at home in our universe.

Part of the effort of *Personal Knowledge* is to create and make acceptable a new idea of the human that is entirely consistent with the concept of evolution. Polanyi tries to persuade us to view ourselves not through ideologies but as beings who have been shaped by evolution. We, therefore, must learn to accept the skills provided by our animal past, including our capability to have personal knowledge.

The structure of *Personal Knowledge*

The book has a spiral structure. Polanyi, switching back and forth between the epistemological and ontological dimensions, progresses in an ever-widening arc. The two dimensions are not independent because from his naturalizing approach; it follows that the observer is the integrated part of nature and reality. Polanyi constantly reflects on this situation. Whenever he reaches some conclusion about knowing, he applies that to the knowing of living beings and humans, which immediately leads to new ontological findings and perspectives. Nonetheless, the main focus is always on knowing.

The book consists of four parts and altogether thirteen chapters. The titles of the parts are: "The art of knowing"; "The tacit component"; "The justification of personal knowledge"; and "Knowing and being." The segmentation function of the parts is not as important as the role of the chapters. The parts do not have their own introductions and conclusions; they merely divide the book into thematic segments. Chapters are much more coherent units; their numbering does not restart in each part but is continuous. An overview of each chapter can also be found at the end of this book.

The most important elements of Polanyi's philosophy

In this part of our preface, we briefly summarize Polanyi's most important thoughts for understanding *Personal Knowledge*, with a special focus on his uniquely novel thoughts.

The new concept of objectivity

Polanyi understands the point of the early Enlightenment as an effort to reduce the subjective in favor of the objective. After the characteristic dogmatism of the Middle Ages, this new scientific attitude proved to be very fruitful. However, the efforts to achieve perfect objectivism are contradictory since they cannot account for humans' capacity for knowing as imperfect persons. Therefore, any epistemology that accepts perfectly non-subjective knowledge as a goal is necessarily unable to explain the actual process of knowing. Such efforts are labeled as "objectivist" by Polanyi, a term we will adopt in this book.

Polanyi contends that we need epistemic principles that reflect and accept the ineliminable involvement of an imperfect person in the process of discovery. This is the program *of Personal Knowledge*, but "personal" here is not the same as "subjective." Instead, personal knowledge aims for "objective" knowledge that is nevertheless grounded in the personal.

The objectivity of statements of personal knowledge derives from them being universal claims, extending beyond the actual person. Such knowledge is always necessarily fallible. Knowledge claims are motivated by the intellectual passions of the person (see below), aimed at grasping reality (but not at final certainty). The act of making a statement is an act of commitment that the statement is true, which gives the semantics of the statement. Any statement, as such, is always a statement of a certain person. The truth of a statement is underpinned by its consequences, the anticipation of consequences, and the intuitions about that statement. If reality, understood in terms of the statement, responds in ever-novel and unanticipated ways, that is a sign of the truth of those statements. But the evaluation and connection to other statements of these developments is necessarily an act of personal knowledge again, driven by objective scientific criteria—but not boundlessly objective.

Fiduciary program

Polanyi's goal is to establish a new epistemic program that is upfront about its own limitations—and less contradictory in comparison with the objectivist approaches described above. Looking for historical connections, he finds St. Augustine, for whom a person's intellectual capabilities relied upon underlying beliefs trustingly shaped in a community of shared beliefs. This close connection between belief and knowledge was severed during the long European secularization. The term "post-critical" in the subtitle of Polanyi's book refers to re-establishing that balance by starting a new era and coming to terms with the ineliminable involvement of improvable elements in knowledge that nevertheless need to be trusted.

The term "fiduciary program" aptly summarizes this approach: through evolution, we have developed the abilities for observing and knowing the world, and also, we have developed trust in those abilities, which is a prerequisite for exercising them. This belief in our abilities is an indispensable part of personal knowledge. We must be able to commit ourselves to the truth of knowledge and act based on that while knowing well that we could be wrong at any time. This is the fiduciary act. It requires giving up the—often hidden—assumption that by removing everything that is subjective from our knowledge, what remains is some infallible core, the rational ego, or the scientific method, etc.

All of this raises the question of what guarantees that our fiduciary acts are not wrong too often, leading to a state where everybody can believe in everything, and our knowledge of reality ceases to develop. Polanyi sees this guarantee *in evolution*. Knowing reality is key to the survival of a species. He explains in detail how each of the high-level epistemic capabilities of the human has been evolved from similar but less developed animal skills. This argument makes it rational to trust our intellectual judgment, especially as there is no better option.

From the post-critical viewpoint, the modern, (boundlessly) critical philosophy or objectivism is just as constraining for science as the dogmatism of the Church was before, only in a different way. The fiduciary program is, therefore, a precondition of freedom, and not only in science but in other dimensions of life as well.

Tacit knowledge

Personal knowledge is a self-reflexive method of knowing in both science and everyday life. Tacit knowledge is part of personal knowledge. Tacit knowledge is something that we cannot fully explicate. Also tacit is the knowledge of animals, which can barely be articulated, if at all. Articulation is really a unique capability of humans; therefore, explicit knowledge expressed in a language distinguishes humanity from the rest of the animal kingdom.

Polanyi does not make a sharp distinction between knowledge and skills. There are reasons for this, as the discussion above of the evolutionary element of the fiduciary program implies. This evolutionary origin is why the skills of a mouse remembering and finding its way in a labyrinth fall under the category of tacit knowledge. The concept of personal knowledge is the basis of an epistemic program that has wide social-cultural consequences as well. The concept of tacit knowledge, however, emphatically illuminates the knowing process from the perspective of a person, including the distinction between two modes of awareness from which one is always tacit due to the structure of knowing.

If we investigate the performance of a person, for example, a knowing act, we will find that there are two kinds of awareness involved. Focal awareness concentrates on the goal of the performance, while the necessary tools, artifacts, and material conditions that we use or exploit for achieving our goals are part of subsidiary awareness. If the focal awareness of the person switches to a subsidiary element, the whole performance falls apart because the respective objects of the two types of consciousness get mixed up. If a cyclist or pianist starts to direct focal awareness to the details of their movements, the performance suffers or falls apart.

Polanyi uses the two types of awareness to explain the use of language, too. Human beings are distinguished in the animal kingdom thanks to the mastery of the complex capacity for articulation, but no knowledge can ever be fully articulated. All knowledge is at least partly tacit because even if very explicitly stated, knowledge is inseparable from the person, from their subsidiary awareness, and their tacit skills. At least when viewed in the context of tacit knowledge, different animals show different levels of development of personhood.

Intellectual passions

The acknowledgment of the role of active intellectual passions follows from the acceptance of the fact that the person is the indispensable locus of knowing. Therefore, the epistemic processes cannot be reduced to impersonal, objective facts. According to Polanyi, the tacit powers that adult humans share with animals and small children (who cannot yet articulate) multiply because of articulation. This is why our thinking is far more advanced than that of most animals, who seem only immediately responsive to events in their environment. Intellectual passions elevated by articulation are part of this human capability. Polanyi distinguishes three kinds of passions by the role they play in science: *selective* passions decide which idea to pursue; *heuristic* passions, which Polanyi connects to creativity; and persuasive passions, which Polanyi links with personal dimensions. These passions are intertwined and work as higher-level operational principles in scientific discovery (see below). A fourth one, *moral* passion, Polanyi elevates and discusses in a different context.

Intellectual passions are present in all acts of knowing. Each time their existence or role is denied by some false epistemology, they are simply disguised by *deceptive substitutions*[1] (see below). Deceptive substitutions hide true motivations, for example, when denying that a phenomenon is interesting for so-called anthropomorphic reasons. Such a denial can be supported by claiming that there are statistical reasons behind the phenomenon's curiosity. This makes it appear that, instead of the work of personal selective passions, there are more objective, exact reasons. Polanyi is convinced that these justifications are always lacking.

Deceptive substitution

The scientist, when trying to justify a certain theory or phenomenon that is interesting and worthy of investigation, masks their true motivations (intellectual passions) with claims about the simplicity, symmetry, and richness of the said theory. These are insufficient accounts of motivations, and they fail to acknowledge the hidden, tacit commitments of the scientist.

So the scientist substitutes their real motivations with concepts better conforming to the objectivist ideal. Such substitutions are connected to the intellectual passions: the scientist pursues a path driven by their intellectual passions, but because these passions are incompatible with the expected methodological requirements, more acceptable reasons are deceptively substituted in place of them.

---

[1] Polanyi uses *pseudo-substitution* at times, apparently with the same meaning.

Moral inversion

Moral inversion is a concept similar to *deceptive substitution.* We may deny the existence of our moral passions, for instance, by accepting a strict materialist worldview, but these moral passions will be at work nevertheless. However, as a consequence, these unacknowledged moral passions will lose their natural place where they can be described and controlled by self-reflection and institutions and become uncontrolled and untamed. At worst, such moral passions will lead to social catastrophes if a fashionable, popular philosophy like Marxism channels those lost moral passions into revolutionary, utopian political and social causes, which in fact are underlain by a crude kind of scientism. In Polanyi's view, the totalitarian political systems of the twentieth century and other dark examples of raw power-wielding are consequences of moral passions becoming unacknowledged and hence uncontrolled due to critical philosophy. This means that Polanyi, albeit indirectly, sees critical, objectivist philosophy as responsible for those cataclysmic events. A successful post-critical philosophy, on the other hand, may lead to a state in which a member of Western culture is in harmony with itself and its cosmic place.

Operational principles and the logic of achievement

Polanyi's application of personal knowledge to living beings and machines creatively develops the concepts of *operational principles* and *the logic of achievement.* His fundamental claim is that machines and living beings—which he acknowledges are machine-like—perform achievements in ways that are not explicable in terms of physics and chemistry. Success in achieving goals defined by normative principles is simply irreducible to the neutral terms of these sciences. This ontological assertion about the nature of machines and living beings is a starting point for the concept of emergence, which explains how these entities come into existence. The nature of machines and living beings is beyond the reach of an analytic perspective that deconstructs everything into its parts; that is, to the standard approach of objectivism and critical philosophy. Machines and living beings and their comprehensive principles can fully be explained only in terms of personal knowledge.

Machines and living beings follow operational principles that are oriented toward achieving goals. Goals and operational principles are not merely physical. But it would be a complete misunderstanding on Polanyi's account to think that goals and operational principles, therefore, transcend the natural. All entities are subject to the laws of physics and chemistry; nonetheless, these laws leave open a degree of freedom in which operational principles become effective. The hierarchy between the two levels is clear: the level of physics and chemistry is below and is more fundamental than the level of operational principles. If the physical-chemical conditions deteriorate, operational principles break down, and a machine fails, or a living being gets sick or dies. But while failure can be explained in terms of physics and chemistry, success cannot. Therefore, Polanyi argues, we need to acknowledge operational principles when explaining success and turn to physics and chemistry to look at the causes of failure; that is, the deteriorated conditions that prevent operational principles from working.

Polanyi's insight affirming that living beings belong to the same class as machines do does not suggest that he takes a mechanistic view of our intellect or even the functioning of the human body. On the contrary, the introduction of operational principles prevents both machines and living beings from reducing to the sum of physical-chemical processes.

Emergence

Polanyi believes that the world has a layered structure. This structure was created *by evolution*, during which higher levels of existence rise from lower levels. Eventually, everything originates from inanimate matter. The concept of *emergence* describes the processes by which higher levels come into existence from lower ones.

Polanyi distinguishes between two types of *perception* in nature. One type of order can be described by the laws of physics and chemistry. For instance, crystallography is about this (lower-level) kind of order: crystals manifest the rules of rightness of crystals. The examples of the other (higher-level) kind of order are equipotentiality in living beings and operational principles in machines. Living beings and machines also manifest rules of rightness; they cannot be described in the neutral terms of physics and chemistry but require personal knowledge.

As we have seen with the logic of achievement, operational principles can only work in the right physical-chemical conditions. In Polanyi's view, the potentiality for a stable, open system that can sustain itself is part of nature. The physical-chemical conditions on Earth were such that they made it possible for the operational principles to become instantiated; hence life emerged from primordial matter. It is important that these principles are not just epistemic concepts but proper features of nature—this is why it is not a category mistake to say, as Polanyi does, that they kick-started life. Polanyi attempts to explain this process in two ways.

Life is the result of a leap. Its predecessor is just lifeless matter. The emergence of human culture is another great leap: when the human species achieved the capacity for articulation, the operational principles of language created a self-sustaining cultural system. The fact that in Polanyi's conceptual system, these two modes of emergence are two similar steps of (emergent) evolution marks

Polanyi's account as a rich and complex account of emergence that is different from many accounts of evolution, which Polanyi criticizes.

Polanyi is committed to the view that both life and culture emerge eventually from matter. Therefore, he is anti-dualist and against any kind of transcendent or supernatural thinking. Instead, he naturalizes religion, which he sees as a system of ethics that is related to other intellectual systems like art and science. He sees statements about God as acts of operating the heuristic intellectual system of religion and not a statement of facts. His views on evolution and the origin of life and humans are incompatible with many popular religious ideas. When he claims that life originates from matter but is not purely material in nature due to the emergence of higher-level operational principles, he attempts to extend natural science to gain proper explanatory power by moving beyond an insufficient materialist account. This new science of nature would include, along with the non-teleological rules of rightness of physics, the teleological operational principles of higher levels as well. This would make natural science consistent with the actual phenomena again. He calls this extended science *ultrabiology*. However, exercising the science of ultrabiology requires the acceptance of the fiduciary program.

Polanyi is one of the most-cited philosophers. His key concepts, especially "tacit knowledge," are routinely employed in several disciplines. However, there is much misunderstanding of his thoughts. These misunderstandings expand as we move from epistemology to his concepts of evolution, emergence, and religion. These dimensions of Polanyi's thoughts are best understood by delving into his biggest achievement, a novel way of viewing the world. Yet, this worldview is hard to describe with simple concepts. At first, we should trust his perspective and absorb the comprehensive meaning of his work: human knowledge is personal. We hope that this book will help to demystify Polanyi's philosophy and provide real access to his key thoughts.

Part One:
The art of knowing

Chapter 1

Objectivity

Goals of the chapter

1. According to Polanyi, the objectivist approach in modern science is false because it cannot correctly identify the reasons for scientific success. Scientists nevertheless try to perfect science; however, most such efforts have merely institutionalized this false approach, which, from the twentieth century, has contributed to the decline of science.
2. The fundamental mistake lies in the meaning attributed to "objectivity." The objectivist approach tries to expel every personal factor from science, believing that these factors are anthropomorphic and only hinder the development of science.
3. Polanyi believes objectivism to be an impossible enterprise. The precondition of the objectivist ideal is a viewpoint that is entirely neutral, not belonging to any particular time and place ("a view from nowhere" or a "God's eye view"). But human beings, because of their unique beings and evolutionary origins, necessarily see the universe from a peculiar point of view, a *center*.
4. With a differing view from that of the history of science, Polanyi shows that tacit skills and intellectual passions play crucial roles in important scientific breakthroughs.
5. Polanyi makes himself clear in that he does not accept materialism because the world does not consist only of material (i.e., physical) parts. He argues that comprehensive, inherent rationality can be discovered in nature.
6. He shows how the absurdity of the objectivist approach is covered up. Although he considers "objectivity" to be highly important in the development of science, his concept of "objectivity" is entirely different than what reigned in much of science and philosophy of science in the first sixty years of the twentieth century.
7. Finally, Polanyi introduces a new concept—*personal knowledge*—that moves the epistemic discussion beyond the subjective-objective dichotomy.

***

Pure, systematic objectivity is unfounded

Polanyi starts his argument with a request to the reader: set aside all convictions and prior notions to imagine what it would mean to be objective with perfect consistency as opposed to the inherent subjectivity found in all individual, human knowledge. In other words, he asks his readers to consider what the literal meaning of pure objective knowledge would be and what the universe and human beings would look like from this purely objective point of view. (Polanyi 1958, 3)

Polanyi proposes the following answers to these questions: since pure objectivity cannot be biased at all and cannot take into account any subjective viewpoints of a human being, from a consistently objective perspective, all small material parts of the whole universe should have been examined *in an absolutely equal manner*. In this case, human beings would not be entitled to any special scientific attention because the solar system and the earth are only tiny parts of the Milky Way, among billions of other stars and planets, and, of course, the Milky Way is just one galaxy among billions of others in the universe. (Polanyi 1958, 3) One of the main problems of pure, systematic objectivity is that it would require that every phenomenon from the set of all possible observable phenomena must be given the same amount of attention.

It does not matter how much scientists talk about the necessity of systematic objectivity. In science, it is clear that no scientist could or would want to contemplate the universe in this manner because it would mean that the earth, a living being, or another person could not appear in scientific knowledge as a particular, interesting entity; we pay attention to these entities only because of our subjective, anthropocentric point of view. Thus, the pursuit of pure or Laplacian objectivity in science is illusory because it ignores the *real situation* in which human knowing is working, as Polanyi notes:

> *"For, as human beings, we must inevitably see the universe from a center lying within ourselves and speak about it in terms of a human language shaped by the exigencies of human intercourse. Any attempt rigorously to eliminate our human perspective from our picture of the world must lead to absurdity."* (Polanyi 1958, 3)

It is important to note that, in his opening section of *Personal Knowledge*, Polanyi immediately addresses the Laplacian ideal of knowledge (see details in ch. 6), which claims that authentic knowledge lacks any human subjectivity, meaning, and point of view and belongs to a super-intelligence that is pure fiction.

BOX 1: Laplacian knowledge

Pierre-Simon de Laplace (1749–1827), a French mathematician and physicist, proposes a super-intellect who, at a given moment, can grasp the place and velocity and any other fundamental characters of all of the objects in the universe. Since this super-intellect is also perfectly familiar with the laws of mechanics, according to Laplace's intention, it can deduce all the possible past and future states of the universe. Later, Laplace's super-intellect was renamed "Laplace's demon" and is frequently mentioned as the real possessor of ideal objective knowledge in various scientific disciplines to this very day. This ideal is often presented as unreachable for human beings and yet worthy of pursuit.

"*We may regard the present state of the universe as the effect of its past and the cause of its future. An intellect which at a certain moment would know all forces that set nature in motion, and all positions of all items of which nature is composed, if this intellect were also vast enough to submit these data to analysis, it would embrace in a single formula the movements of the greatest bodies of the universe and those of the tiniest atom; for such an intellect nothing would be uncertain and the future just like the past would be present before its eyes.*" (Laplace 1902, 4)

Polanyi refers to the "knowledge" of this demon several times to show the inconsistency of the objectivist ideal of knowledge as well as the materialist ontology which follows from it, e.g. (Polanyi 1958, 139-141; 358); or (Polanyi 1959, 48-49). Originally, Laplace's aim was merely to shed light on the meaning of determinism. Only later did this demon become the typical example of the objectivist ideal. It is important to note, however, that Laplace's original wording tacitly already depicts a materialist worldview since it supposes that the universe is but the sum of the (largest and tiniest) material objects and their interactions, and there is nothing else that could impact the past and future movements of these objects. Otherwise, even bright intellect such as Laplace's demon, which is perfectly familiar with the laws of mechanics and the fundamental parameters of the material objects, could not deduce only from these two factors all the possible past and future states of the universe.

The true meaning of objectivity

Polanyi explains the true meaning of "objectivity" with the case of the Copernican revolution. According to some modern writers, the medieval Aristotelian/Ptolemaic geocentric worldview was a sentimental, theologically grounded, anthropocentric account, superseded by the heliocentric worldview informed by a more objective science. Man lost its place at the center of the universe and became one of the many insignificant details with no special place or status. (Polanyi 1958, 3)

However, according to Polanyi, the Copernican revolution was, in fact, a step towards "greater intellectual satisfaction," and, at the same time, it was also a step towards a "more ambitious anthropocentrism." (Polanyi 1958, 5)

> *"In a literal sense, therefore, the new Copernican system was as anthropocentric as the Ptolemaic view, the difference being merely that it preferred to satisfy a different human affection."* (Polanyi 1958, 4)

The Copernican view of the universe is just as far away from a pure, systematic Laplacian "objectivist" viewpoint as the Aristotelian view. Neither the sun nor the earth is anywhere near the center of the universe. This great Copernican leap, in reality, was a step toward a *more abstract* knowledge, which was also more personal. It is legitimate to think of the Copernican system as "more objective" than its predecessor only when a person accepts the shift in what "greater objectivity" means.

> *"It becomes legitimate to regard the Copernican system as more objective than the Ptolemaic only if we accept this very shift in the nature of intellectual satisfaction as the criterion of greater objectivity."* (Polanyi 1958, 4)

Copernicus chose the new heliocentric model because he preferred the *theoretical* heliocentric perspective over our everyday experience that the earth steadily and immovably lies under our feet. Copernicus *had no empirical evidence at all* to support his theory. The clear empirical evidence supports the old Ptolemaic view, but he still commits himself to the heliocentric view because it provides him deeper *intellectual enjoyment* than the old, earthbound one. (Polanyi 1958, 4) It can be stated that the Copernican view is more objective than the previous Aristotelian one; however, it does not reach pure objectivism.

It is important to note that the foundation of this new, Polanyian sense of objectivity is a personal aspiration—*intellectual enjoyment or passion*— which, according to modern interpretations of science, is subjective and thus should be rejected by science. However, for Polanyi, these intellectual passions cannot be subjective; otherwise, they would not be able to lead to more objective scientific theories, not even in a weaker sense (see in detail in ch. 6). Polanyi's goal at this point is only to establish that although there is no ideal knowledge in the sense of pure, systematic "objectivity" (i.e., no God's eye view, the acceptance of the Copernican theory still did not happen only on the basis of subjective factors).

> *"[...the Copernican theory's] excellence is, not a matter of personal taste on our part, but an inherent quality deserving universal acceptance by rational creatures. We abandon the cruder anthropocentrism of our senses—but only in favor of a more ambitious anthropocentrism of our reason."* (Polanyi 1958, 4-5)

According to Polanyi's example, only if people abandon their commitment based on our everyday experience that the earth is in the center of the universe can they accept that the Copernican theory is now satisfactory, not just for us but for every other possible intellectual being in the solar system. (Polanyi 1958, 4)

> *"Since [Copernicus's] picture of the solar system disregards our terrestrial location, it equally commends itself to the inhabitants of Earth, Mars, Venus, or Neptune, provided they share our intellectual values."* (Polanyi 1958, 4)

Thus, Polanyi accepts that the Copernican theory is better, more rational and objective than the previous geocentric one, but he denies that this bright scientific achievement is in any sense the consequence of the Laplacian objectivity of modern interpretations of science. However, a question arises: if the Copernican theory has nothing to do with pure Laplacian objectivity and the empirical observations of that time also refute it, then in what way is it better than the previous theory?

The answer is the following: the new Copernican theory, due to its *inherent rationality,* made contact with reality in new ways, and this will become apparent to the scientific community as the subsequent manifestations of reality are understood within the new framework. In this case, it means Kepler's, Galileo's, and Newton's additional theories, and those new empirical observations which only became possible by the new theory. In the seventeenth century, centuries after Copernicus's death, these additional theories and empirical observations verified the Copernican theory.

> *"We accept [the given theory] in the hope of making contact with reality; so that, being really true, our theory may yet show forth its truth through future centuries in ways undreamed of by its authors. [...] In this wholly indeterminate scope of its true implications lies the deepest sense in which objectivity is attributed to a scientific theory."* (Polanyi 1958, 5)

Although Polanyi breaks with the Laplacian objectivity of modern science, he nevertheless thinks that the Copernican theory is true and more objective than the Ptolemaic account in spite of the fact that, later, important Copernican ideas were disproved by subsequent discoveries (e.g., that the sun is not in the center of the universe at all). Polanyi can do this consistently because, for him, the sign of truth is not that a theory complies exactly with actual empirical observations but that it makes contact with reality in new ways that enable further discoveries and additional theories.

> *"[...] science [has the power] to make contact with reality in nature by recognizing what is rational in nature."* (Polanyi 1958, 6)

It is important to note that those intellectual passions that lead to true theories have to be in contact with reality themselves (see in detail in ch. 6); otherwise, they could not lead scientists to the right path of discoveries. This is also a reason why intellectual passions are not only subjective: they are in clear contact with reality.

Materialism as the consequence of the objectivist ideal of knowledge

It follows that, according to Polanyi, pure, systematic objectivism *entails materialism.* Since objectivism wants to reject the belief that humans and their personal perspectives are special and fundamental in scientific knowing, it has to deny that human beings are different in the ontological sense. Thus, human beings are merely pure material existents just like any other parts of the universe, e.g., the interstellar gas.

However, according to Polanyi, materialism is contradictory since it presupposes a knower who is capable of pure, objective knowing of the whole universe and who, consequently, can*not* be material—otherwise, he would merely be one of the imperfect beings who are able to observe the universe only from a specific point of view (center) that is anchored in their bodies. In chapter 6, Polanyi will elaborate on this claim that materialism is also absurd and covers up its absurdity in the same way as objectivism does.

Thus, Polanyi, in addition to objectivism, also rejects materialism. For him, reality is something more, something hidden that cannot be entirely specified. It will gradually emerge and be revealed for man, for example, with the help of scientific theories. The knowing (and rational) person is an integral part of reality. Furthermore, the goal of science is not only to observe and describe reality in the most exact way based on experience but to understand it more and more deeply, entailing the exploration and explanation of hidden rationality in nature. Thus, according to Polanyi, reality has *both material* and *rational* manifestations.

The formation of the objectivist approach in the history of science

According to Polanyi, there are two important, conflicting traditions in science: the Pythagorean and the Democritean. The former presupposes that numbers are parts of nature; thus, there must be some kind of hidden rationality in reality. The latter emphasizes the material aspects of reality. (Polanyi 1958, 8)

BOX 2: Pythagorean and Democritean traditions in science

Polanyi claims that, from a certain point of view, there are only two major traditions in science in opposition to each other: the Pythagorean and Democritean. His way to differentiate between them is examining whether the given tradition acknowledges or denies that there is inherent, hidden rationality in nature. The exact nature of this hidden rationality plays no role in this categorization. In contrast with this, the common differentiations between traditions of philosophy and the history of science are based on particular ontological and methodological commitments.

According to Pythagoras and his followers, reality itself is nothing but numbers. Plato called himself a Pythagorean in the sense that he acknowledged that reality itself is comprised of hidden rational entities, but he regarded these hidden entities not as natural numbers but as ideas that compose a different, purely rational world. Aristotle changed this theory to the extent that there are not two different worlds. The hidden, inherent, and rational forms are, in fact, in the things themselves along with matter. Matter and forms comprise all earthly existents. Since Polanyi also claims that there is hidden, inherent rationality in nature, he himself is also part of the Pythagorean tradition. However, according to him, this hidden, inherent rationality can be identified not with numbers or ideas but with dynamic *emergent* principles and existents (see ch. 13).

The hidden, inherent rationality of nature was consistently denied for the first time by positivism and materialism. Democritus and his most important atomist follower, Epicurus, obviously cannot be regarded as positivists, and they can only be regarded as materialists in a very broad sense. However, it is a fact that they created a tradition in which rationality is not an inherent part of nature; it merely arises in man at best. That is, it manifests only in certain kinds of complex systems of atoms.

The mechanical worldview which became dominant in modern science is based on the latter tradition. Numbers and mathematics are no longer inherent parts of nature but rational tools, the use of which, nevertheless, remains fundamental for science. (Polanyi 1958, 8) Therefore, reality for science is nothing more than the totality of incidental material objects. These objects, however, are located in a formal and purely theoretical *space-time* that entirely complies with Euclidian geometrics. Thus, the scientific understanding of this space-time, and the laws that determine the motions of material objects in this framework, are based on the intellectual tools of numbers and mathematics. This means that although rationality itself is not an inherent part of nature, thanks to the intellectual powers of scientists, it has still played a pivotal role in science as an objective, Euclidian *background* of nature. (Polanyi 1958, 8)

The discovery of non-Euclidian geometry in the nineteenth century, however, led to the rejection of the reality of mathematics: mathematical propositions became empty tautologies. For many scientists, it meant that there is no true, objective geometry of reality; the different geometries are merely human creations based on arbitrarily chosen axioms.

Consequently, a new *positivist* wave arose by Ernst Mach, and later a new positivist school was formed in Vienna. It rejected every scientific assertion concerning rationality in nature. According to them, scientific theories are only convenient objective summaries of scientific experiences. The last remnants of the Pythagorean tradition had vanished. (Polanyi 1958, 9)

BOX 3: Ernst Mach and positivism

Ernst Mach (1838-1936) was an Austrian physicist and philosopher who had considerable influence on science and the philosophy of science at the end of the nineteenth century and during the first decades of the twentieth century, especially in the German-speaking world. Mach's critique of Newton's theory of space and time clearly impacted Einstein (see the main text), and his positivist concept of science also significantly influenced Niels Bohr and the development of the so-called Copenhagen interpretation of quantum mechanics. In philosophy, Mach can be regarded as the predecessor of the Vienna Circle and the logical positivists.

In Polanyi's interpretation, Ernst Mach was the father of positivism. However, positivism is a much older philosophical tradition that can be traced back even to Pierre-Simon de Laplace. Positivism as an independent school was founded by Auguste Comte (1798-1857), who also created sociology as an independent discipline. However, the chemist Polanyi, in the Austro-Hungarian Empire and in Germany, mainly perceived Mach's influence who, as opposed to Comte, was working in science and in these disciplines in particular.

In the first half of the twentieth century, positivism became the main philosophical tradition in science. According to Polanyi, positivism is clearly objectivist as it tries to eliminate traditional metaphysics both in science and philosophy, which, of course, means the rejection of any kind of hidden, inherent rationality in nature.

According to the positivist interpretation of the history of science, Albert Einstein did nothing more than give an answer for the negative outcome of the Michelson-Morley experiments, which could not detect the movements of Earth in contrast to the ether.

BOX 4: The Michelson-Morley experiments

Albert Michelson and Edward Morley, two American physicists, performed their famous experiment in 1887, which was later repeated several times by Morley and D. C. Miller with much more precise instruments. Originally, the aim of the experiment was to define the velocity of Earth as compared to the ether.

At that time, light was considered to be a wave that spread through the absolutely motionless ether. According to this theory, since the earth is orbiting the sun, it is not in absolute rest. If the velocity of the light compared to Earth could be measured in perpendicular directions, the results should show a slight difference due to the revolving motion of the earth in the absolutely motionless ether. However, the measured difference was significantly less than they expected.

By the traditional interpretation, this negative result led to the rejection of the concept of ether. It also led to Einstein's theory of special relativity, according to which the speed of light is absolute, while the motions of other objects are relative. Therefore, the results of the Michelson-Morley experiment have to be the same in every direction. However, the situation is not as simple, as Polanyi himself points out, because the Dutch physicist Hendrik Lorentz's dynamical ether theory explains the results just as clearly as Einstein's theory of special relativity. It follows that the rejection of the concept of the ether is not merely based on the experimental results.

Furthermore, the shift happened despite the fact that it led to severe theoretical problems. According to Einstein, light is both a particle and wave at the same time, which is a clear conceptual contradiction. Similarly, light is supposed to be a wave that has no medium to travel in, also a severe contradiction.

Moreover, the experimental results, which were repeated over one hundred thousand times during the next three decades, mainly by D. C. Miller, do not show that Earth is motionless as compared to the ether (that is, the speed of light is an absolute constant). Rather, his results demonstrate that the speed difference is significantly smaller than it has to be due to the revolving motion of the earth around the sun. Therefore, the results of the experiments, in fact, did not verify Einstein's concept that the speed of light is the same constant in every inertial frame of reference.

The results of D. C. Miller's experiments, as Polanyi emphasizes, were ignored by the scientific community without any well-established explanation. (It is worth noting that there is no well-established explanation for D. C. Miller's experimental results to this very day; the scientific community simply regards them as measurement errors.)

In Mach's view, Newton's theory of space-time based on Euclidian geometry is meaningless, silly metaphysics that has to be expelled from science because it has nothing to do with real, observable phenomena. Polanyi, however, shows that this is not the case at all. He even consults Einstein himself via an intermediary. Einstein acknowledges that the Michelson-Morley experiments had little importance in the discovery of the theory of relativity. His thinking was led by different intuitive speculations concerning the nature of light. (Polanyi 1958, 10)

BOX 5: Polanyi's question to Einstein

Polanyi, via one of their mutual acquaintances, N. Balazs, asked Einstein what the role of the Michelson-Morley experiments in the development of his theory of relativity was. The answer was the following, according to N. Balazs's account:

"*Today I discussed with Einstein the basic ideas which have led to the foundation of the special theory of relativity.*
The result is about the following:
There were basically two problems whose contemplation was of fundamental importance. (1) The problem he is referring to in his autobiographical sketch about the impressions of an observer moving with the velocity of light and viewing a lightwave; (2) the lack of symmetry of action between ψ current elements and ψ magnets. [In the pre-relativistic electrodynamics of moving media, it made a lot of difference whether you move a wire with a current relative to a magnet, or the magnet relative to the wire.] (1) suggested to him that the velocity of light must play a privileged role; (2) seemed strange since, among other reasons, he felt that the situation is to be determined by the relative velocities which are the same. I hope I do not misrepresent him. The Michelson-Morley experiment had no role in the foundation of the theory. He got acquainted with it while reading Lorentz's paper about the theory of this experiment (he of course does not remember exactly when, though prior to his papers), but it had no further influence on Einstein's considerations, and the theory of relativity was not founded to explain its outcome at all." (Polanyi 1958, 10-11)

Similarly to Einstein, his followers accepted the theory of relativity not because of objective experimental evidence. Although D. C. Miller's results seriously contradicted Einstein's theory for a while (Polanyi 1958, 13), it was still accepted because of the intellectual beauty and rationality of the theory. (Polanyi 1958, 14)

According to Polanyi, Einstein did not show that Newton's concept of space-time is meaningless, as Mach thought. It showed that it is *false.* Einstein did this *by intuitively grasping the deeper meaning of rationality in nature.* Newton's concept of space-time is, in fact, based not just on Euclidian geometry but on the nature of ordinary human experiences where the objects are observed before a strict material background in every case. However, Einstein supposed that, for the material world as a whole, there is no condition in which the fundamental material entities can appear, e.g., light in the ether. This means that Einstein, at least tacitly, supposed that ordinary human experiences that were adapted to earthly conditions should be ignored in the cosmic scale while investigating the question of the movement of light; just as Copernicus supposed that ordinary everyday human experience should be ignored when considering the question of the real cosmic movements of planets. Instead, Einstein based his intuitive approach on the deeper rationality of nature. (Polanyi 1958, 12)

Einstein started to use non-Euclidian mathematics to describe the changes in time and space generated by physical objects, which, according to Polanyi, was a significant step towards the Pythagorean tradition. By doing so, Einstein corrected two errors: one in the thinking of Newtonian physicists; and one in that of Machian physicists. The followers of Newton still tried to understand the deeper rationality hidden in nature by Euclidian geometry, which is strongly based on ordinary human experiences, while the positivist followers of Mach tried to fix this mistake by rejecting all kinds of rationality in nature. Einstein's balanced approach highlighted the problems in both approaches. (Polanyi 1958, 16)

It is worth mentioning that Polanyi cites Kepler in detail to show how obvious his mystical and passionate commitment to a heliocentric worldview was. This level of passion, in contrast to the positivist picture, perfectly complies with the true nature of science. (Polanyi 1958, 7)

BOX 6: Polanyi's quotation of Kepler

The following quotation comes from Kepler's *Harmonices Mundi* (*The Harmony of the World*) Book V, ch. 10 and Preface.

"Of what sort vision is in the sun, what are its eyes, or what other impulse it has [...] even without eyes [...] for judging the harmonies of the [celestial] motions, [...] for those inhabiting the earth; not easy to conjecture [...] in the sun there dwells an intellect simple, intellectual fire or mind, whatever it may be, the fountain of all harmony."

"What I prophesied two-and-twenty years ago, as soon as I discovered the five solids among the heavenly orbits—what I firmly believed long before I had seen Ptolemy's Harmonics—what I had promised my friends in the title of this fifth book, which I named before I was sure of my discovery—what sixteen years ago I urged to be sought—that for which I have devoted the best part of my life to astronomical contemplations, for which I joined Tycho Brahe [...] at last I have brought it to light, and recognized its truth beyond all my hopes [...] So now since eighteen months ago the dawn, three months ago the proper light of day, and indeed a very few days ago the pure sun itself of the most [marvelous] contemplation has shone forth—nothing holds me; I will indulge my sacred fury; I will taunt mankind with the candid confession that I have stolen the golden vases of the Egyptians, in order to build of them a tabernacle to my God, far indeed from the bounds of Egypt. If you forgive me, I shall rejoice; if you are angry, I shall bear it: the dice is cast, the book is written, whether to be read now or by posterity I care not; it may wait a hundred years for its reader, if God himself has waited six thousand years for a man to contemplate His work." (Polanyi 1958, 7)

Polanyi notes that Kepler "even went so far as to write down the tune of each planet in musical notation." (Polanyi 1958, 7)

Deceptive substitutions

An important question arises because if, according to Polanyi, the objectivist approach is not just false but absurd and, in reality, nobody uses it in scientific practice, then what is the reason that this fact is not clear for everybody?

Polanyi's answer is that we use "*deceptive substitutions*" in scientific language by which we conceal the real meaning of certain scientific assertions. Polanyi does not give details at this point; rather, he just gives a few examples to clarify what the point of these deceptive substitutions is. His more detailed example is *the principle of simplicity* on the basis of which one theory is preferred in contrast to another. However, according to Polanyi, the real meaning of simplicity can neither be defined nor understood merely by objective criteria if we do not refer to the concealed rationality that guides our choice and cannot be defined explicitly. The theories of relativity and quantum mechanics are not simple at all at their explicit mathematical level unless we tacitly refer to their inherent scientific rationality. (Polanyi 1958, 16)

It is also the case with the concept of *scientific beauty* that was mentioned several times by Einstein's followers to explain why they chose Einstein's theory over H. Lorentz's or over D. C. Miller's experimental results. The concept of beauty, in reality, refers to that inherent rationality of nature that was demonstrated in Einstein's theory. Using another term instead of rationality, however, we conceal this fact as if it were only about some aesthetic point so that we do not contradict the positivist dogma that there is no deeper, inherent rationality in nature.

Acknowledging the fact that there is no real objective perspective or knowledge and yet preferring one theory over another based on which one is closer to this ideal is also a deceptive substitution. In this case, scientists are just concealing that the basis of their choice remains the same absurd objectivist ideal despite the supposedly bright self-reflection used to justify it. However, the real reason that it is also a deceptive substitution is that the real purpose of this scientific talk is to conceal the fact that partly incidental personal skills and intellectual passions are the basis of how scientific choices are actually made.

Conclusion

The modern interpretations of science are based on a sharp distinction between subjectivity and objectivity. Polanyi, however, claims that this dichotomy is false. Personal senses, skills, intuitions, and intellectual passions that are connected to the inherent rationality of nature are not mere subjective obstacles one has to overcome but integral parts of science. These skills and passions guide the scientists to reveal and understand the different orders, laws, and rationality of nature. Polanyi calls these intellectual skills and their workings *personal knowledge*. This bridges the deep abyss between the two sides of the subjective-objective dichotomy and leads the scientist from the subjective perspective to real objective knowledge. (Polanyi 1958, 17)

Chapter 2

Probability

Goals of the chapter

1. Polanyi shows that statements concerning reality cannot be made without trust, belief, and conviction, that is, according to the requirements of the objectivist approach of science.

   "The purpose of this book is to show that complete objectivity as usually attributed to the exact sciences is a delusion and is in fact a false ideal. But I shall not try to repudiate strict objectivity as an ideal without offering a substitute, which I believe to be more worthy of intelligent allegiance; this I have called 'personal knowledge.'" (Polanyi 1958, 18)

2. Polanyi shows that scientific measurement is based on the appraisal and evaluation of errors; consequently, dealing with probability is an integral part of it.
3. Then he shows that probability statements need heuristic, personal judgments; that is, they do not comply with the impersonal ideal of objective knowledge. Together with the previous point, this means that measurement, the bedrock of empirical science, does not meet the requirements of the objectivist ideal.
4. Polanyi also shows that the evaluation of anomalies; that is, such scientific results that do not comply with the actual theories contradict the objectivist ideal too.
5. He examines the nature of statements concerning probability and differentiates between non-personal probability statements (e.g., "The probability that I'm going to cast a six is 1/6.") and such statements that refer to such events, correlations, laws, etc., that occur or apply only with some probability (e.g., "I'm going to cast a six."). These latter statements are not probability statements; it is only that certain probability can be assigned to them; in other words, one can state them only with a certain degree of personal confidence.
6. Polanyi shows that the heuristic selection and proposal of scientific hypotheses necessarily belong to this second category. If the hypotheses were not proposed in this manner, then scientists had to look for

evidence considering an unlimited number of logically equal hypotheses. This, according to Polanyi, is the parody of scientific practice.

7. It is necessary for a proposal of a scientific hypothesis to be more than just a probability statement, even in the case when the scientist cannot present the evidence by which they made their choice.
8. Polanyi claims on the basis of his examination of the nature of probability statements that a personal *fiduciary act* by which someone commits himself to the content of a statement is an integral part of that statement. This is true for all statements, not just statements concerning probability. This will be elaborated in ch. 8.

***

It is important to note that the title of the chapter does not cover the whole content of the chapter because Polanyi goes further in the examination of the general nature of statements. Also, the question of probability along with randomness and order will be raised again in the next chapter.

Unambiguous and probability statements

Polanyi differentiates between two kinds of statements with regards to ambiguity: *unambiguous* and *probability* statements. We can see from his examples that unambiguous statements are such statements that are called universal statements in the philosophy of science. In the case of probability statements, there is no way to assert the statement with certainty, only with some probability; that is, with some degree of confidence. In science, both of their goals are the same: to express knowledge *in an objective, non-personal way.*

Unambiguous statements

Polanyi's examples for unambiguous statements are the formulas of classical mechanics. In principle, any possible future and past states of an object can be calculated by a few exact formulas if the physical parameters of the actual state of the given object are known exactly. However, this knowledge can only be regarded as purely objective *if someone does not take into account the application of mechanical formulas to empirical facts.*

> *"The derivation of data and checking of data that bridge the gap between our instrument readings and the magnitudes figuring in our formulae can never be fully automatic. For any correlation between a measured number introduced into an exact theory and the corresponding*

> *instrument readings, rests on an estimate of observational errors which cannot be definitively prescribed by rule."* (Polanyi 1958, 19)

Scientists have no readily available and perfect sensory and knowing skills (see in detail in ch. 4). The instruments by which they generate data that is turned into exact formulas are also imperfect. Random measurement errors occur regularly, which have to be corrected with some method of appraisal. In other words: there are only probability statements referring to these errors; there is no unambiguous, ever-valid rule or method of correction for every case. (Polanyi 1958, 19)

Also, scientists have to rely on their personal judgments in cases with contradicting results found during the verification of a scientific theory; and it is the normal way of how science is done. (Polanyi 1958, 20)

Probability statements

In contrast with classical mechanics, in most of the sciences, only probability statements are available. Polanyi's examples are quantum mechanics and one of the experiments of Charles Darwin. In these cases, there is no possibility to express scientific knowledge in unambiguous ways only with some probability. This means that there is no exactly known data in the equations of quantum mechanics, unlike in the case of classical mechanics. For example, the place of an object is only known by the probabilities of it being in one of its possible positions.

> *"Probability statements can never be strictly contradicted by experience, even if we assume that all external perturbations and all observational errors are entirely eliminated."* (Polanyi 1958, 21)

This means that in the case of a theory in which there are probability statements, the personal evaluation of the scientific experience is always necessary, even if one does not take the problem of errors and measurements into account. (Polanyi 1958, 21)

So, according to Polanyi, probability statements are useful for science not because they objectively synthesize the scientific data as positivism claims, but because they give room for scientists to evaluate the theories personally and compare them to measurement data.

> *"[...] if no strictly objective restriction can be derived from the assignment of this probability, we may expect to find in it instead some guidance to our personal participation in the event to which the probability statement refers."* (Polanyi 1958, 21)

Polanyi explains this with dice rolling. In the strict sense, no experience can contradict the following statement: "the probability of a double six is 1/36" since it does not refer to future events, only to future *possibilities*; so in a purely objective sense, *it cannot be in conflict with any experience.* Even if someone, for example, cast three double sixes in a row, the probability of which is only 1/46656, neither this rare event refutes nor questions the original probability statement (and it is, of course, true for any other series).

The fact that the series of three double sixes is surprising does not sign a conflict between experience and the probability statement, but it is the expression of personal participation and evaluation which are connected to random events—*this and only this personal participation establishes a relation between experience and the probability statement.*

> *"I ascribe universal validity to my appraisals of probability, in spite of the fact that they make no predictions which could be contradicted by any conceivable events."* (Polanyi 1958, 22)

Therefore, since by strict objective criteria, it cannot be decided which events are too improbable to treat as real—it has to be decided on the basis of *personal evaluation.* (Polanyi 1958, 23) There is not, however, any objective rule as to when this has to happen or what is the exact probability when an event has to be regarded as not real. (Polanyi 1958, 26)

The situation was exactly the same when, for example, Darwin wanted to determine if there is a real difference between the heights of plants in cases of cross-fertilization and self-fertilization or the observed differences are merely the consequences of other random variations.

Of course, it is both possible and necessary in science to create formal rules and methods for this kind of scientific problem. But these formal rules and methods never replace but merely *support* the personal participation and evaluation of the results. In Polanyi's words:

> "*But such formalization is likely to go too far unless it acknowledges in advance* that it must remain within a framework of personal judgment." (Polanyi 1958, 29)

If a scientist, following an objectivist conceptual framework, thinks that these formal rules and methods are sufficient to answer a scientific question of this kind, then it is simply a deceptive substitution (see ch. 1). Since, in practice, they still evaluate the results of the formal methods on the basis of their personal intellectual skills and participation, they reflect on their work by formalisms only to cover their personal participation in the results and to comply with the objectivist concept of knowledge. Polanyi explains this kind of deceptive substitution with the following example:

According to R. A. Fischer, a statistical method can be given concerning Darwin's above-mentioned experiment by which it can be shown that "the probability that cross-fertilization increases the growth is more than 95%." This explicit sentence, however, means nothing in itself without any personal judgment because it has no connection with experience. It is only a probability statement referring to a certain possibility. The non-personal, objective *probability* of this statement is *not the same* as the *content* of a statement *based on personal conviction*, for example, that "cross-fertilization increases the growth"—it only numerically expresses the degree of one's personal conviction concerning the given statement. (Polanyi 1958, 25) So it has no real meaning because no experience can contradict it. The reason for this is that the whole statement is, in fact, the following: "The probability that cross-fertilization increases the growth is more than 95% and the probability that it reduces it is less than 5%." The statement is perfectly objective and non-personal. But it has nothing to say about reality; therefore, in itself, it does not correspond to any scientific practice. (Polanyi 1958, 24)

So the real scientific statement, which, of course, was meant by Darwin himself too, is that cross-fertilization *increases* the growth of plants. Experience can contradict this belief, and this is exactly the source of its meaning. It is not an unambiguous statement of classical mechanics without exceptions, only a biological statement that was verified by the statistically analyzed results of Darwin's experiment. Scientists have to judge based on their personal intellectual skills which results can be accepted as evidence during this process and which cannot; they have no other option. The conviction of a scientist about the truth of the statement follows from this process (science in practice). The probability of the statement refers to the degree of his conviction which, however, cannot be identified with a non-personal probability statement. (Polanyi 1958, 25)

Therefore, the rephrasing of the real scientific statement—that cross-fertilization *increases* the growth—into an objective, non-personal statement—the probability that cross-fertilization increases the growth is more than 95%—in itself is only a *deceptive substitution* concealing the personal judgment of the scientist. This statement makes no real claim concerning reality; it just contains a possibility, but, of course, in practice, everybody would understand it in its original sense—that cross-fertilization *increases* growth. And everybody has to understand it this way because non-personal rephrasing means nothing. It follows that an assertion of a statement based on a sincere conviction is more than its pure content. (Polanyi 1958, 27)

> *"An unasserted sentence is no better than an unsigned cheque; just paper and ink without power or meaning."* (Polanyi 1958, 28)

> *"[...] no sincere assertion of fact is essentially unaccompanied by feelings of intellectual satisfaction or of a persuasive desire and a sense of personal responsibility."* (Polanyi 1958, 27)

It is worth noting that, according to Polanyi, the meaning of a statement is established in the relationship between reality and the person who asserts it. An explicit sentence in itself means nothing; it only has content. It only gains any meaning if, and only if, a person has committed itself to its content as something referring to reality. Polanyi, however, will only detail this claim in ch. 8.

Conclusion

By the detailed examination of probability, Polanyi has shown that heuristic decision-making is necessary in science. It *cannot be depersonalized* and can *only partially* be formalized. Scientific measurement and data collection are already based on personal evaluation of probabilities. According to this claim, the degree of a scientist's conviction towards a scientific statement is not equal to a non-personal probability statement. Substituting a scientific statement based on real personal conviction with a non-personal probability statement makes it as meaningless as the pure, systematic objectivist description of the universe is (see ch. 1). Polanyi generalizes this to not only probabilistic but all kinds of assertions: a statement becomes meaningful if and only if scientists commit themselves to it and assert it on the basis of real personal convictions.

Chapter 3

Order

Goals of the chapter

1. Following the train of thoughts of the previous chapter, Polanyi wants to show that the presumption of randomness is possible only by the prior assumption of order. This is the main function of the "Wales" example.

2. Polanyi claims that the recognition of order in nature is an achievement by two related personal judgments.

   "*Every kind of human knowing, ranging from perception to scientific observation, includes an appreciation both of order contrasted to randomness and of the degree of this order.*" (Polanyi 1958, 38)

   a. *The recognition of order:* we can distinguish order from randomness. In other words, we can find explanations for comprehensive order in nature, suggesting that they are not merely random and illusory but the consequence of some natural law or ordering principle.

   b. *The appraisal of the depth of an order:* we can appraise the extent and depth of an order. By this second personal skill, we cannot just explain the recognized comprehensive order in nature, but we can appraise the position of the given order on the scale of coherent existence.

3. Polanyi harshly criticizes the neo-Darwinian theory because it attempts to explain the comprehensive order of life with the concept of randomness[1].

***

The recognition of order in contrast to randomness

> "*When I say that an event is governed by chance, I deny that it is governed by order […].*" (Polanyi 1958, 33)

---

[1] For further discussion see Paksi, Dániel. "The Meaning of Randomness: The Laplacian Fault of neo-Darwinians According to Michael Polanyi". *Appraisal: The Journal of the British Personalist Forum* 10: 3, 43-53, 2015.

This means that randomness is nothing else but a statement of a correlation concerning nature. A recognized comprehensive pattern is either a consequence of order—that is, some kind of ordering principle—or only randomly occurring—that is, there is no ordering principle. In the second case, we call the comprehensive but meaningless pattern random, although it is, in fact, the consequence of entirely deterministic processes—it is just that they are not part of any comprehensive pattern. According to Polanyi, the randomness of different processes can be interpreted only in relation to an orderly pattern. (Polanyi 1958, 33-34)

To clarify the problem, Polanyi uses a thought experiment that is based on his personal experiences.

> *"At the border between England and Wales you pass a small town called Abergele. Its railway station has a beautifully kept garden in which, sprawling across the lawn, you are faced with the inscription, set out in small white pebbles: 'Welcome to Wales by British Railways.' No one will fail to recognize this as an orderly pattern, deliberately contrived by a thoughtful station-master."* (Polanyi 1958, 33)

According to the thought experiment, if someone finds a garden with pebbles, there are three options:

1. There is no recognizable pattern in the pebbles of the garden.
2. There is a recognizable, comprehensive pattern in the pebbles—a Wales inscription—arranged intentionally by the stationmaster.
3. There is a recognizable, comprehensive pattern in the pebbles—a Wales inscription—that is not the consequence of some intentional human activity but only the material processes in the garden (e.g., of a storm).

Suppose that, according to Polanyi's little story, someone notices the Wales scheme in the garden. At once, he would think that it is case two above, so it was arranged intentionally by the stationmaster, and it is a meaningful pattern of order. How could he know for sure that this is the case and the Wales scheme is not just the consequence of randomness?

We are inclined to answer this question by referring to the extremely small possibility that could lead to the Wales scheme without any human interference. It can be expressed numerically by relating the particular arrangement of the pebbles of the Wales scheme to all possible arrangements of the pebbles in the garden (one particular pebble arrangement to x possible arrangements; that is, 1/x), which would be a very small number. Therefore, the Wales scheme was intentionally arranged by the stationmaster (case 2) because the probability of the other case (case 3) is awfully small.

However, this answer is not so simple and unambiguous. On the contrary, it is highly problematic because the probability of any particular arrangement of pebbles in the garden, including those not forming a recognizable pattern (case 1), would be exactly the *same* (one to all possible arrangements; that is, 1/x). Nevertheless, contrary to this fact, no one would ever claim that an unrecognizable arrangement of pebbles is also the intentional work of the stationmaster. Moreover, no one would ever claim that it is the consequence of randomness as claimed when the Wales scheme was not arranged intentionally (case 3) although the probabilities are the same; *both of them were arranged by the same unintentional physical processes in the garden.*

So the questions are the following: why does someone believe, when he spots the Wales scheme, that it complies with case 2, and why does he not notice that his argument for this belief based on probability calculations is inadequate?

> *"We have assumed from the start that the arrangement of the pebbles which formed an intelligible set of words appropriate to the occasion represented a distinctive pattern. It was only in view of this orderliness that the question could be asked at all whether the orderliness was accidental or not. When the pebbles are scattered irregularly over the whole available area they possess no pattern and therefore the question whether the orderly pattern is accidental or not cannot rise."* (Polanyi 1958, 34)

The answer to the first question is that we involuntarily regard the Wales scheme as unique compared to other pebble arrangements (case 1). This means that the basis of our judgment is our *personal knowledge* concerning letterings, gardens, and railway stations. And the answer to the second question is that we are arguing inadequately because, according to the concept of objectivism, we want to give a *non-personal* answer to the original question; that is, why we think that the Wales scheme was arranged by the stationmaster intentionally (case 2). To achieve this non-personal answer, we try to determine the "*a contrario*" probability of our statement about the origin of the Wales lettering; that is, the probability of the Wales scheme not having been arranged by the stationmaster. However, our argument was deceptive not just because the value of probability is the same in all possible arrangements but also because it was based on our personal knowledge, that is, on the recognition of the Wales scheme. Since if there is no recognizable scheme (case 1), the question that it is the consequence of randomness (case 3) or someone's intentional act (case 2), in fact, could not arise at all—thus, numerically expressed probabilities also cannot be stated relating to it.

It is worth noting that it counts as another case of deceptive substitution (see ch. 1). if someone, according to the objectivist ideal, conceals their personal participation by a seemingly objective and non-personal formal argument.

The reason that probabilities cannot be assigned to case 1 is that it is known, with 100% exact probability, that the arrangement of pebbles in the garden is the consequence of deterministic physical processes; and it is, of course, true for every other possible arrangement of case 1. A value of probability can be assigned to any possible pebble arrangement of case 1 *only if* it is "*a contrario*" contrasted to a peculiar possible arrangement, for example, to the Wales scheme. In itself, every possible arrangement is absolutely determined either by physical processes or by the actions of the stationmaster.

Suppose a new case 2b when there is a recognizable, comprehensive Wales pattern in the pebbles arranged intentionally by the stationmaster, but the arriving person does not recognize it because, for example, he is Chinese and has never heard of the Latin alphabet. According to the situation, probabilities cannot be assigned to the arrangement as in case 1. This, on the one hand, shows the consequences of the lack of personal knowledge without which there is no question about probabilities at all. On the other hand, this shows that *the arrangement of pebbles in a Wales scheme* is not to be identified with the *Wales lettering* itself.

To explain this, let's consider the similarities and differences in the three original cases. In case 1, the particular arrangement of pebbles is just as much the consequence of deterministic physical processes as the Wales scheme of case 3. In case 1, no one would ever say that the arrangement of pebbles is random, but in case 3, they would say exactly that. The contradiction, however, is only illusory since, in case 3, the real meaning of the statement is not that the arrangement of pebbles is random but that *the recognized Wales scheme is random.* That is, the statement that the arrangement of pebbles in the garden is the consequence of the lower-level deterministic processes is, in fact, *valid in both cases.* It is a vacant statement because it was known with certainty (100% probability). The Wales scheme is random because, in contrast to case 2, there is no *specific order or ordering principle behind the particular arrangement of pebbles.* Therefore, for the presumption of randomness and thus for probability calculations, the *prior* assumption of ordering principles is needed.

> *"I wish to suggest that the conception of events governed by chance implies a reference to orderly patterns which such events can simulate only by coincidence."* (Polanyi 1958, 36)

It does not mean, of course, that all recognized patterns are real; one can easily be wrong and commit a *subjective mistake.* Polanyi's examples are the constellations by which the horoscopes calculate: astrology assumes that the particular order of constellations is meaningful. But this assumption is false because they are merely the consequences of deterministic, physical movements of stars, just like the Wales scheme in case 3 is the consequence of deterministic, physical movements of pebbles in the garden. This means that the creators of horoscopes believe that they can grasp a working of a hidden ordering principle in the constellations. However, they are wrong because, in reality, the meaningless, deterministic movements of stars can display the wonderful patterns of constellation *only by chance.* (Polanyi 1958, 36)

Polanyi's critique against the "*a contrario*" method of R. A. Fischer (see in ch. 2) is based on this recognition, and it is also the basis of his argument against neo-Darwinism (see below and in detail in ch. 13).

BOX 7: R. A. Fischer and Darwinism

Sir Roland Fischer (1890-1962) was a British statistician who played a significant role in the first half of the twentieth century to develop population genetics and ground the neo-Darwinian synthesis. At the beginning of the twentieth century, it was not clear at all what is the relation between Charles Darwin's theory of natural selection and genetics based on Gregor Mendel's principles. Darwin's theory, due to observation, simply supposed that there are variations of living beings among which selection works and does not explain the formation of variations. Mendel, at the same time, explicated the principles of heredity entirely independently from the theory of natural selection 1. Furthermore, concerning its methodology, Darwin's work is rather a historical explanation by detailed empirical examinations, while genetics is highly theoretical and mathematized. It is R. A. Fischer's merit that this methodological gap was bridged, and the Darwinian mechanism of natural selection was reinterpreted by the principles of genetics. According to Polanyi, by this interpretation, Fischer moved the Darwinian theory to the direction of objectivism, and this process only continued and strengthened in the neo-Darwinian synthesis.

> *"To test the probability of such coincidences and hence the permissibility of assuming that they have taken place, is the method of Sir Ronald Fisher for establishing a contrario the reality of an orderly pattern."* (Polanyi 1958, 36)

That is, for the sake of illusory objectivity, to allow the possibility that astrological patterns can arise from stellar constellations in ways other than by chance or that a storm can create a Wales scheme in ways that are not random. This means that one does not openly undertake its personal judgment, according to which an orderly pattern is real and another one is not. Instead, according to the concept of objectivism, assigns to them non-personal probability values by which they are categorized (see ch. 2). Consequently, this method allows that an illusory scheme might be the consequence of a real ordering principle because it renders probability value to this logical possibility. And, vice versa, it allows the possibility that a real pattern is only the consequence of chance.

However, the question as to whether a pattern of order is meaningful and real cannot be answered by the probability of stellar and pebble movements. This is because our probability statements do not refer to an exact star or pebble arrangement *but to* existent or nonexistent *ordering principles* of recognized patterns. As we have seen in cases 1 and 3, the appraisal of an orderly pattern cannot be done only by the examination of lower-level deterministic processes. (Polanyi 1958, 36)

> *"On these grounds I suggest, quite generally, that the appraisal of order is an act of personal knowledge, exactly as is the assessment of probability to which it is allied."* (Polanyi 1958, 36)

Polanyi emphasizes that the fact that all meaningful knowledge is personal does not mean just a new label on the concept of knowledge since "the degree of our personal participation varies greatly within our various acts of knowing." (Polanyi 1958, 36) According to this, one can speak about:

1. *Objective facts* in the case of the arrangements of pebbles, or of the casting of three double sixes, or of the movements of stars;
2. *Personal facts,* if a meaningful pattern is arranged in the pebbles, or if one appraises the casting of three double sixes as an astounding random event;

3. *Subjective mistakes,* if someone, such as the astrologists, appraises the constellations as real, meaningful patterns, or if he thinks that a real, meaningful pattern of order is aroused by mere chance.

All three types are based on personal knowledge, but they have greatly different meaning in reality. (Polanyi 1958, 36-37)

If, through a detailed examination of the lower-level deterministic processes, it cannot be determined which scheme is real and which is only an illusion, then another principle is needed in order to do this task:

> *"[...] man has the power to establish real patterns in nature, the reality of which is manifested by the fact that their future implications extend indefinitely beyond the experience which they were originally known to control. The appraisal of such order is made with universal intent and conveys indeed a claim to an unlimited range of as yet unspecifiable true intimations."* (Polanyi 1958, 37)

Polanyi hints at such concepts, which will be detailed in later chapters. For the time being, he can establish confidently that human beings have the knowledge and skills to recognize and appraise real patterns in nature. That is, they have the power not just to see the meaningless and deterministic lower-level processes to which in themselves even probabilities could not be assigned, but to recognize the meaningful higher-level patterns too. Regarding these patterns, however, unambiguous statements cannot be formulated—only probability ones. Therefore, the possibility of subjective mistakes is always present as someone might make false judgments about the meanings of these statements. However, if one is right, then the future consequences will confirm that in unforeseeable ways. (Polanyi 1958, 37)

The appraisal of the deepness of an order

> *"Statements of probability can be made about random systems and about significantly ordered systems in so far as these are affected by interaction with random systems."* (Polanyi 1958, 37)

> *"Probability statements are therefore always based on an anterior knowledge of randomness."* (Polanyi 1958, 38)

This interaction enables the recognition of meaningful—that is, orderly—patterns in nature or meaningless—that is, random—ones. Behind the former, there are some ordering principles; behind the latter, there are not (see table 3.1).

Table 3.1: *The relation of random and ordered systems to lower-level random systems*

Random, meaningless system	**Orderly, meaningful system**
The W-scheme of Cassiopeia	Wales lettering
↕ Determining interaction	↕ Random interaction
Movements of stars	Movements of pebbles
There is no ordering principle	There is an ordering principle

> *"Randomness [that is, the movements of stars and pebbles] alone can never produce a significant pattern, for it consists in the absence of any such pattern; and we must not treat the configuration of a random event as a significant pattern, whether by attributing to it fictitiously a distinctiveness that it does not possess, as in the case of the scattered pebbles, or by granting it erroneously a specious significance, such as the fulfillment of a horoscope."* (Polanyi 1958, 37-38)

Of course, this does not mean that the lower-level background processes—that is, the movements of stars and pebbles—are random in themselves, only that they are random in relation to recognized schemes (either orderly or random themselves).

> *"Any entity—whether an object or determinate process—will be the more clearly set off against its background, the more amply its internal particulars show steadiness and regularity—combined with an amply confirmed absence of any co-variance between these particulars and those of the background."* (Polanyi 1958, 38)

In case 1 of the Wales example, there is a perfect correspondence between a "pattern" and the background—and as a consequence, there cannot be any recognizable pattern.

In case 2, there is no correspondence between the meaningful pattern and the background at all since the stationmaster moves the parts of the pattern, and the parts of the background are moved by the physical processes in the garden. However, if the stationmaster stops arranging the pebbles, then the movements of the pebbles are going to disrupt this real pattern too.

In case 3, a random pattern is separated from its background to some extent, but in reality, it has no meaning at all—although we are inclined to think that. It can be seen that it has no meaning, from the fact that it shows minimal permanence and regularity depending on how much the shifts of its parts hinder the recognition of the false pattern.

According to Polanyi:

> *"We may even grade the intensity of coherent existence on this scale. Owing to its more significant internal structure a human being is a more substantial entity than a pebble. The difference can be appreciated by comparing the sciences of anatomy and physiology with the range of interest offered by the structure of a particular type of pebble. Every kind of human knowing, ranging from perception to scientific observation, includes an appreciation both of order contrasted to randomness and of the degree of this order."* (Polanyi 1958, 38)

However, he explains in detail only in ch. 13 why he thinks a human being is higher on the scale of coherent existence than a pebble.

> *"An ordering principle can be extrinsic, as in the case of a message or any other artifact, or intrinsic, as shown in the ordered coherence of a solid body and in any stable configuration, whether static or dynamic."* (Polanyi 1958, 39)

Case 2—when the Wales pattern was arranged by the stationmaster—is an example of extrinsic ordering principles. Polanyi gives an example for the intrinsic ones too. With this example—which is a thought experiment—he argues again his claim that randomness can never create a meaningful pattern because its essence is precisely the fact that it lacks every such pattern. (Polanyi 1958, 37) In Polanyi's thought experiment, there are ideal dice on an ideal plane that is shaken by random Brownian motion. (Polanyi 1958, 39)

In experiment 1, due to an extrinsic ordering principle, all dice show one on top. An adequately strong and permanent Brownian motion will disrupt this orderly pattern, and a maximum disorder will be created concerning the faces of the dice.

In experiment 2, according to an extrinsic ordering principle, all dice show one on top, but they are biased in favor of six. First, the random Brownian motion will also disrupt the pattern, but then it will release the operation of the intrinsic ordering principle, and at the end of the process, most dice will show six on top in a state of equilibrium.

In experiment 3, most dice show six on top, and they are biased in favor of it, so actually, this is precisely the equilibrium state of experiment 2. Now let's strengthen the random Brownian motion significantly; consequently, this new orderly pattern will also be destroyed, and a maximum disorder will be created again, as it was at the end of experiment 1.

> *"Experiment 2 shows that random impulses may release the operation of forces which tend to produce a stable pattern. Where such a dynamic ordering principle is lacking, as in Experiment 1, the existing order is destroyed in the long run even by the weakest random impulses. But random impulses of a sufficient strength as applied in Experiment 3 will destroy likewise any dynamically stable order [...]"* (Polanyi 1958, 39-40)

It follows that a higher-level orderly pattern is not in the least the consequence of lower-level random processes that just release the operation of ordering principles present in the situation. At the same time, lower-level random processes have to operate within certain boundaries; otherwise, no order will be created. So the creation of an orderly pattern has two preconditions: 1) an *ordering principle* of that kind of pattern, and 2) *adequate lower-level random processes.* Without any of them, there would be no real orderly pattern.

The critique of the neo-Darwinian theory

The above is precisely the reason why Polanyi harshly criticizes the neo-Darwinian theory, which claims that living beings are the consequences of random mutations and random environmental conditions by natural selection. That is, it claims that the meaningful, orderly patterns of life—i.e., living beings themselves—are the consequences of lower-level random processes. However, since the lower-level processes in themselves, of course, are deterministic, such a claim can *merely be stated in relation to* higher-level orderly patterns; therefore, *it has to be presupposed,* at least tacitly, before the claim, that there are higher-level orderly patterns of life, which in this case means the coherent existence of living beings. (Polanyi 1958, 35)

> *"[...] we must conclude therefore that the assumption of an accidental formation of the living species is a logical muddle. It appears to be a piece of equivocation, unconsciously prompted by the urge to avoid facing the problem set to us by the fact that the universe has given birth to these curious beings, including people like ourselves."* (Polanyi 1958, 35)

However, he explains his critique in detail only in ch. 13.

BOX 8: Polanyi's relation to Darwin

In most cases, Polanyi speaks simply about Darwinism. It is only evident from the context which kind or level of Darwinism he refers to exactly. To understand Polanyi's critique, it is enough to see the difference between Darwinism and neo-Darwinism.

The theory of natural selection is the foundation of all kinds of Darwinism. From a scientific point of view, the most important difference between the Darwinian and the neo-Darwinian theories is that the former does not include genetics yet (see above in the R. A. Fischer BOX). But from a philosophical point of view, there is a more serious conflict since Darwin himself does not claim that the theory of natural selection can explain the phenomena of evolution in nature and thus that natural selection is the only fundamental mechanism in evolution. However, this is perhaps the most important principle of neo-Darwinism. Darwin argues more modestly, only claiming that natural selection can explain the creation of new species in nature in the same way as artificial selection; that is, livestock breeding can explain the development of new subspecies of dogs, pigeons, sheep, etc.

It follows that when Polanyi argues against explanations by genetic mutations or when he states that the theory of natural selection as a sole fundamental mechanism cannot explain any real evolutionary features of nature (see the main text), then he is, in fact, not even contradicting Darwin, because Darwin himself never claimed these things. So Polanyi does not debate Darwin at all—merely the neo-Darwinian expansion of his theory. Moreover, if we do not identify Darwinism with neo-Darwinism, then Polanyi could be regarded as a Darwinian because, in the limited Darwinian sense, he completely accepts Darwin's theory of natural selection (see ch. 13).

An example of the workings of personal knowledge in science

Polanyi explicates two detailed examples of the workings of personal knowledge from his own scientific field. The complicated one is about crystallography, and the more telling one is about the law of chemical proportions; that is, the exact proportion of different elements in chemical compounds. Of course, the process of measurements cannot be formalized entirely, but here at least, there are some methods to filter out the errors.

However, in the case of chemical compounds, there is no rule that the proportions of elements have to be as simple as 1:1 or 1:2 and such. As a matter of fact, scientists would have obtained *much more objective* proportions if they had not supposed that the proportions have to be simple. From this point of view, their choice could seem arbitrary; that is, according to the concept of objectivity, they had to insist on such results as 23/45 and did not simplify them by subjective, arbitrary reasons—in this case, to 1/2. However, this insistence on the objectivist ideal could have inhibited the development of science because Dalton's atomic theory was based on the arbitrary concept of simple chemical proportions. Later, Bohr and Rutherford only continued his work to explain the deeper meaning of this—at first sight—arbitrary system of proportions. (Polanyi 1958, 40-43)

> *"Dalton's discovery of the atomic theory was itself based on the evidence of simple chemical proportions and thus confirmed the intimations of reality contained in the appraisal of this orderly pattern."* (Polanyi 1958, 42)

This means that the chemists of those ages, by ignoring objectivity and exactness based on their personal scientific skills and knowledge, and assuming simple fractions for the proportions of elements in chemical compounds, appraised rightly that behind the results of their experiments lies a hidden ordering principle and not just randomness.

Conclusion

Polanyi has argued that, with the help of personal knowledge, one can recognize order in relation to randomness and can also appraise the extent and depth of that order. The latter is done by correlating the intrinsic stability of the order against the random processes of the background. On this basis, one can explicate probability statements concerning natural laws and ordering principles behind the recognized order.

However, Polanyi only details the system of ordering principles in Part Four of his book. In his words:

> *"For the moment, it is enough to recognize here that, in affirming these fundamental laws of nature, we accredit our capacity for knowing randomness from order in nature and that this distinction cannot be based on considerations of numerical probabilities, since the calculus of probabilities presupposes, on the contrary, our capacity to understand and recognize randomness in nature."* (Polanyi 1958, 40)

Chapter 4

Skills

Goals of the chapter

1. Polanyi draws attention to the unspecified parts of skills by detailed examination of their structure and inner workings.

2. He regards skills (swimming, cycling, playing the piano, various crafts, cotton spinning, and the use of industrial equipment) and scientific knowing as the same in terms of their unspecifiability. So, according to him, the most well-known example of tacit knowledge, bicycle riding, is not just an example of the ways of practical skills but scientific ones, too.

   "*Science is operated by the skill of the scientist and it is through the exercise of his skill that he shapes his scientific knowledge. We may grasp, therefore, the nature of the scientist's personal participation by examining the structure of skills.*" (Polanyi 1958, 49)

3. By the destructive analysis of skills, he shows that there are not just accidental, historic examples of unspecifiability (e.g., Mesmer and his followers do not know what principle lies behind hypnosis), but that it is the necessary part of any achievement. This is referred to by the concept of *logical unspecifiability.*

4. From this, Polanyi concludes that there are two kinds of awareness: *focal* and *subsidiary.* During a performance (e.g., playing music), the artist pays attention focally to the piece, while he is aware of the parts (e.g., pressing a key) of his acts only subsidiarily. As soon as he turns his attention to the parts, the performance will fall apart.

5. Because of the unspecifiability that is interwoven in personal knowledge, commitment is a necessary part of it. Although the parts of personal knowledge are necessarily unspecified, one can still rely on personal knowledge by commitment.

6. This is also the reason why transfer and acquisition of knowledge are limited to personal relationships based on trust and why it "tends to survive in closely circumscribed local traditions." (Polanyi 1958, 53)

7. So Polanyi shows that personal knowledge has deep bodily and social—that is, biological and cultural—evolutionary roots.

8. He claims that the real meaning of objectivity is that a scientist commits himself to a *universal vision of reality.* The degree of this commitment necessarily extends beyond the person and his understanding. However, methods such as verification, falsification, decisive experiments, etc., together with a false concept of objectivity, could disburden the scientist from this serious personal responsibility. Consequently, exact sciences may appear as the sum of formulas that refer to experience, but, in reality, the scientist still relies on his (now disguised) personal conviction in every assertion about such relations, and he can only explicate the exact formulas this way. (Polanyi 1958, 49)

***

The rules in learning an art

Human skills are guided by different *rules.* For example, for those who want to swim, it is necessary to learn the appropriate breathing technique and to keep more air in their lungs during swimming than usual. If one wants to ride a bicycle, it is necessary to learn how to turn the handlebar with small, careful movements in order to master balancing. These principles are not generally known and in themselves have no significance at all—they become meaningful only during the performance of the given skill. (Polanyi 1958, 49) Polanyi's detailed examples are swimming, cycling, and playing the piano. (Polanyi 1958, 49-51)

Of course, it is possible to describe through exact mechanical equations how a falling cyclist regains balance—it would have significant scientific value—but during cycling, it has no use; nobody can ride a bicycle by calculating mechanical equations in their mind. On the contrary, the complicated mechanical equations would likely divert the person's attention from the task at hand, which would be disruptive. If the pianist turns his attention to the pressing of keys rather than the piece as a whole, then his performance will fall apart. (Polanyi 1958, 51) The reason for this is that attention has only *one focus.* However, the artist is aware of several different things because awareness has *multiple scopes.*

Focal and subsidiary awareness

Polanyi differentiates between two kinds of awareness: *focal* and *subsidiary.* The two awarenesses are mutually exclusive; that is, one pays attention to something either focally or subsidiarily. (Polanyi 1958, 55)

During an act of skill, the act itself is in the focus of attention (focal awareness), so the act of cycling, the act of swimming, the playing of a piece, etc., itself. At the same time, a person, through this act as a whole, is aware of every tool, tangible part, and rule by which the act is accomplished (subsidiary awareness); he does not pay attention to these parts but is subsidiarily aware of the bicycle as a tool, of his muscular movements as tangible parts, and the rules to keep more air in the lungs or to regain the balance through small, careful movements, etc. (Polanyi 1958, 55)

Contrary to the rules of act, the other subsidiary parts and tools can be determined fairly well by turning the focus of attention on them. However, as we have seen, the act itself will then stop, and a new *knowing act* will start. If someone does not know an act of skill as a whole, they will never be able to learn that skill through only the detailed analysis of the parts—which is another act. This is the reason, according to Polanyi, that an act cannot be logically specified by its parts. Also, the learning of an act is a tacit practical process and not a theoretical deduction from the explicitly described parts. (Polanyi 1958, 56)

BOX 9: Tacit integration

Since the nature of the process by which the subsidiary parts form a focal comprehensive whole is tacit, Polanyi calls it *tacit integration*—but he establishes it in detail only in *Logic and Psychology*. According to this, the structure of skills can be depicted like this:

Subsidiary parts → Tacit integration → Focal whole

This means that it cannot be specified by formal logic. Formal logic creates explicit deductive relations among focal parts, but during a comprehensive act, one is aware of the parts only subsidiarily. This is the reason comprehensive wholes cannot be deduced from focally explicated parts; moreover, as we have seen, destructive analysis based on formal logic and explicitly determined parts tear down comprehensive acts as such into pieces. So this type of process has a necessarily tacit nature because one is aware of the parts only subsidiarily; therefore, the act has to be integrated into a comprehensive whole *tacitly*: here, there is no possibility for explicit deductive inferences.

"*This difference between a deduction and an integration lies in the fact that deduction connects two focal items, the premises and consequents, while integration makes subsidiary bear on focus.*" Logic and Psychology 32

So the difference between deductive inference and tacit integration is the following:

Subsidiary parts → Tacit integration → Focal whole

Focal part(s) → Explicit deduction → Focal part(s)

In *Personal Knowledge*, Polanyi merely provides an example to explain tacit integration, according to which, if one is not subsidiarily aware of a tool during an act but pays attention focally to it during a knowing process, then it could not be regarded as a tool, merely as an object. An object becomes a tool, not because of its material existence but because one trusts it during an act to reach one's end goal. (Polanyi 1958, 56)

"*In this sense I should say that an object is transformed into a tool by a purposive effort envisaging an operational field in respect of which the object guided by our efforts shall function as an extension of our body. My reliance on it for some end makes an object into a tool, even though it may not achieve that end.*" (Polanyi 1958, 60)

Destructive analysis

Destructive analysis, according to Polanyi, is a method of knowing the critical tradition that tries to understand and explain the meaning and workings of a given phenomenon or act by the detailed examination of their focal parts. Therefore, the point of this method is focusing on the parts and tearing down the given phenomenon or act into the smallest possible parts. (Polanyi 1958, 51)

Polanyi asserts that this method is essential in science. On the one hand, it is useful to demonstrate the falsehood of certain phenomena, skills, and methods; and on the other, the detailed knowledge of parts could deepen the understanding of the given phenomenon and strengthen the given skill. (Polanyi 1958, 51) At the same time, he also draws attention to the dangers of the method because an excessive belief in it could easily lead to the false concept that everything can be deduced from the material parts, which, according to Polanyi, is a total misunderstanding of the real workings of science. And if a new scientific skill or discovery—for example, mesmerism—is conceptually misunderstood by its discoverers, then the systematic demonstration of the conceptual contradictions of the interpretation of the new knowledge could even lead to it being rejected, despite the fact that it is clearly working.

BOX 10: Mesmer and hypnosis

Franz Anton Mesmer (1734-1815) was a German physician who, during the second half of the eighteenth century, was the first to use extensively and successfully the technique that we now know as hypnosis. Today, hypnosis is a well-known, scientifically recognized phenomenon, although the exact causes and processes behind it are not well established.

Mesmer believed that hypnosis was related to the magnetism in the human body, and he named it "animal magnetism." Scientifically, it was an important step forward because, in contrast to the former mystical and religious explanations of the phenomenon, he tried to interpret it by *scientific concepts.* This interpretation supposed such concepts as, for example, that the tide-like processes of the body are the consequences of the movements of the sun and the moon—he tried to apply Newton's theory—which is known to be false today.

During hypnosis, Mesmer moved magnetic objects over and placed them on patients' bodies; in addition, sometimes, the patients had to drink liquids containing different magnetic powders to enable the healing process through the body's animal magnetism. Besides these methods, the treatments included some simple elements of hypnotic suggestion, but Mesmer regarded these as accompaniments to the phenomena of animal magnetism.

The scientific community accused Mesmer of being a "quack," and he even had to flee Vienna following one particular case.

Polanyi interprets the story of mesmerization as follows (he refers to John Elliotson (1791-1868), who was one of the most important of his followers):

"The critics of Mesmer and later of Elliotson found it easy to demonstrate that the manipulations which these men said they were performing were in themselves ineffectual. Elliotson had expounded a whole system of laws governing the alleged transmission of animal magnetism. He claimed that the magnetism of a glass of water, the drinking of which caused cataleptic trance, could be graded by dipping one finger into it, or two fingers, or the whole hand. Another 'law' declared that mucous surfaces of the subject, like those of the tongue or the eyeball, were capable of receiving a greater mesmeric stimulus than the skin. Later Elliotson announced that gold and nickel were more sensitive to mesmeric influences than base metals

like lead. All this was nonsense and was easily proved to be nonsense. And since the assumption had not yet dawned upon anyone that hypnotic suggestion was the effective agent of Mesmerism, the conclusion seemed inevitable that Elliotson's subjects were impostors, who were either deluding him or colluding with him. In vain did Elliotson bitterly appeal: 'I have given details of 76 painless operations in the name of common sense and humanity, what more is wanted?' Not until the concept of hypnosis was established as a framework for the facts, could these facts be eventually admitted to be true." (Polanyi 1958, 51-52)

Personal and objective facts

On these grounds, Polanyi differentiates between *personal* and *objective* facts. (See this from another point of view in ch. 3.) All knowledge referring to a comprehensive, meaningful act or phenomenon is a personal fact. This means that, for example, the recognition that someone is riding a bicycle is a personal fact because it does not refer to subsidiary, precisely specifiable tangible parts but to an integrated, comprehensive act. It can then be decomposed by destructive analysis into tangible parts, which, according to their relatively more objective nature, can be called objective facts, e.g., the concrete muscular and neurological movements of the body during cycling. On the one hand, this means that personal facts are the *preconditions* of destructive analyzes and more objective facts. On the other hand, there are no absolute objective facts: even the elementary particles of quantum mechanics do not seem to be entirely specifiable. (Polanyi 1958, 63)

Tangible and intellectual tools

Polanyi determines tools not by their structural parts and material compositions but their roles in human acts. This connects the concepts of body and tool strongly.

> *"Our appreciation of the externality of objects lying outside our body, in contrast to parts of our own body, relies on our subsidiary awareness of processes within our body."* (Polanyi 1958, 59)

According to him, an external object is focally observed and localized during a knowing act. If something is not observed in this way, it is not considered an object. Nonetheless, it could be entirely real and play an important subsidiary role in an act.

A person's body observed focally is also an object. However, normally, no one considers his body in this way and is aware of it only subsidiarily. This means that a person's body is normally used as an *internal tool* to reach different goals; that is, from this point of view, it is used just as any other *external tool*—for example, a hammer—to get the nail into the wall.

> *"Our subsidiary awareness of tools and probes can be regarded now as the act of making them form a part of our own body. [...] While we rely on a tool or a probe, these are not handled as external objects. [...] We pour ourselves out into them and assimilate them as parts of our own existence. We accept them existentially by dwelling in them."* (Polanyi 1958, 59)

BOX 11: Embodiment

The phenomenon Polanyi speaks about is referred to as "embodiment" or "extended cognition" in the philosophical literature. This means that the mind extends in the body and even beyond the body into external objects. Mind, contrary to the traditional Cartesian concept, is not another extensionless substance; while the body is an extended material substance, it works in the body and through the body, and it can even extend beyond the boundary of it. It follows that the mind cannot identify with the body as materialism imagines, which, in reality, means that the mind does not exist.

Polanyi gives a new interpretation to this phenomenon of embodiment with his concepts of focal and subsidiary awareness since the concept of subsidiary awareness allows a new explanation about how different bodily or external parts become parts of the mind while retaining their own nature.

Human beings, of course, do not just have hammer-like tangible tools but also intellectual tools that play fundamental roles in scientific skills.

> *"Hammers and probes can be replaced by intellectual tools; think of any interpretative framework and particularly of the formalism of the exact sciences."* (Polanyi 1958, 59)

Someone is aware of words during an act, e.g., a talk, in exactly the same way they are aware of a hammer when the nail is being hit into the wall—that is, subsidiarily. Of course, words are useful in other ways than a hammer. They are used to getting informed or getting in touch with others. The way we use them, however, is the same. Since scientific knowing is mainly based on human language and its conceptual categories, it is as important in science as in any peculiar scientific method or theory.

> *"Our language includes the numerals and the elements of geometry, and it refers in these terms to laws of nature whence we can pass on to the roots of these laws in scientific observations and experiments."* (Polanyi 1958, 59)

> *"Like the tool, the sign or the symbol can be conceived as such only in the eyes of a person who relies on them to achieve or to signify something."* (Polanyi 1958, 61)

According to what we have seen, table 4.1 categorizes our tools.

Table 4.1: *Tangible and intellectual tools*

	Internal	**External**
Tangible	Body	Artificial tools
Intellectual	Spoken language	Written language

However, written language cannot be entirely classified into only one category since every written language is also an external artificial tool. Of course, every written language is, at least partly, a spoken language. Furthermore, it is important to note that in a sense, every spoken language is an extended version of body signs; that is, unarticulated voices and body language (see ch. 5), as artificial tools were originally made to be the extension of the human body. This means that every highly explicit and sophisticated scientific method and practice is *ultimately based on deep bodily skills, sensibility, and passions,* which Polanyi discusses in detail in the next chapters (mainly in ch. 5, 6, and 8). In the twentieth century, human technology and scientific knowledge allowed the connection of spoken languages with artificial tools and the organized installation of artificial tools in the human body.

Existential, representative, and denotative meaning

Polanyi differentiates among three different kinds of meaning: *existential, representative,* and *denotative.* (Polanyi 1958, 58)

Every kind of order has existential meaning. This means that material parts exist in a comprehensive system with its own ordering principles and meanings. For example, in the case of cycling, the ordering principles are the rules of the act of cycling, and the existential meaning of cycling is that somebody is riding a bicycle (see in detail in ch. 8). In the case of a frog, the ultimate ordering principles are the ordering principles of life and evolution (see ch. 13), and the primary existential meaning of the frog is that it lives and acts according to its own nature and goals.

An order made by human beings—for example, a portrait—generally has some kind of representative meaning, too; that is, it conveys a message. If this message is articulated by language, then the order also has a denotative meaning; e.g., the Wales lettering of the previous chapter is not just a pile of pebbles that have their own particular existential meaning but also conveys a concrete intellectual content for the arrived passengers.

The role of trust in personal acts

To use an intellectual tool successfully, one has to trust that it is just as capable of doing its task as a hammer is capable of getting a nail into a wall. If we believe that the given tool or concept is not capable of doing the task or solving the problem in question, then we will simply not use it. Trust is necessary because, during the act, the tool is in a subsidiary position of awareness and is only a tiny part of the comprehensive human performance, just as any part of the body. In other words, the acting person is not aware of it focally. Without trust, the (tangible or intellectual) tool is simply an alien object—and if someone pays attention to it focally, it could become an object of critical examination. The main precondition of the critical method and destructive analysis is that one has to back away a few steps to examine the object focally without trust. (Polanyi 1958, 60)

Therefore, a language as an internal intellectual tool has to be trusted; that is, if someone uses a language, he cannot doubt it, but *he has to accept it without criticism.* This means, according to Polanyi, the assimilation of the conceptual and explanatory framework of that language. Intellectual tools can be used as tools only by trust and without criticism in a subsidiary position of awareness. (Polanyi 1958, 60)

The tacit nature of the real fundaments of scientific beliefs

According to Polanyi:

> *"[...] the supposed pre-suppositions of science are so futile because the actual foundations of our scientific beliefs cannot be asserted at all.*

> *When we accept a certain set of pre-suppositions and use them as our interpretative framework, we may be said to dwell in them as we do in our own body. Their uncritical acceptance for the time being consists in a process of assimilation by which we identify ourselves with them. They are not asserted and cannot be asserted, for assertion can be made only within a framework with which we have identified ourselves for the time being; as they are themselves our ultimate framework, they are essentially inarticulable."* (Polanyi 1958, 60)

That is, the real foundations of scientific beliefs and theories cannot be asserted because a conceptual and explanatory framework in which any assertion can be made is a necessary precondition for such an assertion, and one should assert the basic conceptual and explanatory system of one's scientific beliefs.

So the real foundations of scientific beliefs and theories are inarticulate; that is, they are tacit intellectual tools that are accepted uncritically and internalized and embodied in a subsidiary position of awareness. Yet, if someone partly explicates them during a critical examination, they will not find them convincing at all. The reason for this is that, according to the nature of any critical examination, as we saw in the previous point, one has to make them external, put them in a focal position of awareness, and become alienated from them. The real foundations of scientific beliefs and theories, however, gain their meaning only in practice in a subsidiary position of awareness, just as any other rules for practicing human skills. (Polanyi 1958, 61)

The workings of craftsmanship

The rules of practicing an art also fundamentally determine the workings of craftsmanship. On the one hand, this means that they can be examined focally to some extent, and on the other that, as we have seen, they gain their real meaning only in practice; therefore, craftsmanship itself can be acquired solely through practice.

Practice once more excludes the critical method and destructive analysis; that is, for the sake of acquiring a new craft, one has to uncritically accept both the unarticulated rules of art for that craft and the peculiar situation in which that craft can be learned by example and practices. It follows that craftsmanship can be acquired only in such institutions where the apprentice accepts the authority of their master and trusts and follows the institution's traditions unconditionally. If the apprentice does not do this but tries to use critical methods against their master and institute, or tries to learn the rules of the art of that craftsmanship by destructive analysis when perhaps even the master cannot explain those rules perfectly, then that apprentice will learn nothing and will soon end up being excluded from the institution.

> *"These hidden rules can be assimilated only by a person who surrenders himself to that extent uncritically to the imitation of another. A society which wants to preserve a fund of personal knowledge must submit to tradition."* (Polanyi 1962: 53)

Since science is a kind of craftsmanship, this all holds true for science, too. (Polanyi 1958, 49)

So skills are based on such tangible and intellectual tools, fundamental conceptual and explanatory systems, and rules of the art of which one is aware only subsidiarily. These can only partially be articulated focally, gaining their real meaning only in practice. Therefore, skills can be learned solely in traditional institutions and with trust.

> *"[…] the possession by humanity of an immense mental domain, not only of knowledge but of manners, of laws and of the many different arts which man knows how to use, comply with, enjoy or live by, without specifiably knowing their contents."* (Polanyi 1958, 62)

Conclusion

According to Polanyi, personal knowledge based on skills can work only if one acknowledges both his bodily and social reality uncritically, because this is the root of the (tangible and intellectual, internal and external) tools and rules of art which, after assimilation and embodiment, allow him to reach his goal.

> *"It is the act of commitment in its full structure that saves personal knowledge from being merely subjective. Intellectual commitment is a responsible decision, in submission to the compelling claims of what in good conscience I conceive to be true. It is an act of hope, striving to fulfill an obligation within a personal situation for which I am not responsible and which therefore determines my calling. This hope and this obligation are expressed in the universal intent of personal knowledge."* (Polanyi 1958, 65)

Polanyi details the act of commitment in ch. 10, but here he has shown the essence of it. To use an object as a tool, one has to trust it and assimilate it in his person in the sense of subsidiary awareness—it is the act of commitment itself. This is true, of course, for any intellectual tool and rule of art, too. We internalize and assimilate the language, words, and concepts as intellectual tools, and laws and morals as rules of activities, and commit ourselves to them uncritically.

> *"This reliance is a personal commitment which is involved in all acts of intelligence by which we integrate some things subsidiarily to the center of our focal attention.*
> *Every act of personal assimilation by which we make a thing form an extension of ourselves through our subsidiary awareness of it, is a commitment of ourselves; a manner of disposing of ourselves."* (Polanyi 1958, 61)

Part Two:
The tacit component

Chapter 5

Articulation

Goals of the chapter

1. Polanyi introduces the tacit roots of language and language-based thinking, emphasizing that everything that is not articulated and not explicit is tacit.
2. He builds his argument on the most important empirical results of animal tests regarding learning, concluding that, upon setting aside our language-based knowledge, the remaining intellectual capabilities of man would hardly exceed the tacit, prearticulation intellectual capabilities of primates.
3. Thus, there is only one tiny tacit advantage on the side of man, which, if leaned on, grants an enormous intellectual advantage to animals by the acquisition of articulated language. To put it in another way: there is no considerable leap of (biological) evolution among humans and advanced animals.
4. Polanyi emphasizes that articulation ignites a self-invigorating process that extends those tacit skills and capabilities which made the process of articulation possible.
5. Also, he shows that articulation is never complete; the lingual intelligence of man is still based on tacit skills during both the use of language and language-based conceptual thinking.
6. The undeniable presence of tacit elements provides another argument proving that the concepts of objectivity and truth criticized in the previous chapters are unacceptable; thus, we need to reinterpret these concepts.

***

"Admittedly the scientist's art of knowing, which I have surveyed previously, is on a higher level than the child's or the animal's and can be acquired only in conjunction with a knowledge of science as a formal discipline. Other intellectual skills of a high order are acquired similarly in the course of a continued formal education; and indeed our mute abilities keep growing in the very exercise of our articulate powers." (Polanyi 1958, 70)

Basic types of animal learning

Polanyi sorts the forms of learning into three groups. The first two are more primitive, while the third is formed by connecting the first two. (Polanyi 1958, 71-74)

Type A is *trick learning*, introduced by B. F. Skinner's experiment with rats, in which the rat substantively learned a trick to get food by depressing a lever. This type of learning involves obtaining a particular movement technique to reach a specific purpose. (Polanyi 1958, 71-72)

Type B is *sign learning*. Polanyi illustrates this with a rat experiment too. During the experiment, the rat gradually learns which of the two doors with different markings has food behind it by recognizing the sign that indicates the food beyond the door. This type of learning also requires acquiring a movement technique, but this technique is not necessarily new. The essence of purpose-driven learning here is rooted primarily in the achievement of *perception*. Every animal possesses the ability to understand things coherently. Sign learning is the extension of perceptive ability by intelligence. (Polanyi 1958, 72-73)

Type C learning is *latent learning*. It is also exemplified with a rat experiment. Polanyi calls it latent learning when a rat understands, e.g., the structure of a labyrinth, by forming a *mental scheme* of it, which he is also able to use in practice. From its latent knowledge of the structure of the labyrinth, the rat is able to conduct particular types of behavior, which can be seen as a rudimentary logical inference. Latent learning is the source of the ability of problem-solving, which is discussed in detail later in this chapter. (Polanyi 1958, 73-74)

Since each of the three types of learning is prearticulational, all of them are *tacit*. This leads us to conclude that, according to Polanyi, *tacit logic* also exists alongside explicit logic (see ch. 4).

In each of the three cases, learning can be divided into three separate sections. The first is the *irreversible, heuristic* process of learning, and the second is the *reversible, routine-like* application of all that is learned. Case A is about recognition and then repeating a trick; in case B, observation happens first, followed by an iterative reaction to a sign, while in case C, understanding gives birth to a posterior solution for a routine problem. (Polanyi 1958, 76)

In the case of humans, the fundamental types of learning manifest themselves in invention (type A), observation (type B), and interpretation (type C). In the sphere of highest intellectual abilities, we can discern the definitive presence of type A in engineering, type B in natural sciences that build on inductive inferences, and type C in exact sciences applying deductive inferences (see table 5.1).

Table 5.1: *Different types of human and animal abilities*

	Type A	Type B	Type C
Basic animal ability	Trick learning	Sign learning	Latent learning
Human ability	Kant	Observation	Interpretation
Highly differentiated human ability	Engineering	Natural sciences	Exact sciences

Polanyi's interpretation is somewhat superficial since the correspondence of different animal, and human types of learning and intellectual abilities are not as simple as he depicts. However, in the train of thought, it is of no serious relevance, as the point of Polanyi's statements is that the antecedents of human intellectual abilities can be found in the tacit skills of animals defining the potential of articulated intellectual capabilities of man.

Types of lingual utterances

Polanyi distinguishes three types of lingual utterances. These are as follows: 1. *expressions of feelings*; 2. *appeals to other persons*; 3. *statements of fact.* (Polanyi 1958, 77) By analyzing the latter, he points out that:

> *"[...] the ingredient of personal passion inherent in and necessary to even the least personal forms of speech which my argument seeks to exhibit."* (Polanyi 1958, 77-78)

The operational principles of language

Upon examining the operational principles of language, Polanyi differentiates two of them. The first controls the process of linguistic *representation*; the second the *operation of symbols* to assist the thought process. (Polanyi 1958, 78)

The first operational principle of language is the *Law of Poverty*, implemented with the Laws of Grammar, Iteration, and Consistency. Language—compared to the number of thoughts ever conceived—uses only a small number of symbols. Thus, lingual utterances often have multiple or indefinite meanings. The meaning of spoken sentences could only be identified in an exact manner if there would be an independent and unique symbol replacing every single spoken sentence. In this case, however, exact and "univocal" symbols would have no meaning at all, since the meaning of a word is formed and manifested by its repeated usage when, upon the occurrence of unique situations, the factual personal experience is tacitly attached to a word of the language. Thus, every single word has to occur relatively often, which means that language has to be poor regarding the number of words. (Polanyi 1958, 79)

To express original situations, we have to combine these relatively often occurring words that, in turn, presume the existence of certain rules and principles to be considered upon attaching these words side by side. During consequent usage, words have to recur identifiably. Therefore, the existence of grammar becomes a necessity.

> *"Since the world, like a kaleidoscope, never exactly repeats any previous situation [...], we can achieve consistency only by identifying manifestly different situations in respect to some particular feature, and this requires a series of personal judgments."* (Polanyi 1958, 79-80)

This means that we have to be able to distinguish relevant phenomena from the meaningless, random background. In each case, it is a personal decision based on generalization and trust that by using a specific word, we manage to capture a meaningful phenomenon. Consequently, by the use of language, we inevitably sort things into different classes: there will be red and black, animate and inanimate, light and heavy, etc., things.

> *"These expectations form a theory of the universe, which we keep testing continuously as we go on talking about things. So long as we feel that our language classifies things well, we remain satisfied that it is right and we continue to accept the theory of the universe implied in our language as true."* (Polanyi 1958, 80)

Polanyi compares the operation of language to chemical compounds. Chemical elements are like words, and chemical bonds are like grammar, building up a theory of the universe that we consider to be true. In the case of descriptive sciences, the link between reality and language is even tighter, as Polanyi explains it thoroughly later in ch. 12.

The second operational principle of language is the *Law of Manageability*. Churches and pyramids are symbols but not of manageable size.

> *"In the most general terms, the principle of manageability consists in devising a representation of experience which reveals new aspects of it."* (Polanyi 1958, 81-82)

By "manageability of symbols," Polanyi means that, with the help of these symbols, our experiences can be reorganized, our thoughts can be reconsidered again and again from the outside, and as a result of that, we can obtain new knowledge. This ability grants an enormous advantage for man over animals. He illustrates the relevance of symbolic reorganizing of experiences by giving the example of a map that provides several correlations of this nature that are not available when simply looking at pure geographical data. (Polanyi 1958, 82)

> *"To speak is to contrive signs, to observe their fitness, and to interpret their alternative relations; though the animal possesses each of these three faculties, he cannot combine them."* (Polanyi 1958, 82)

The use of language is a skill or craft since applying words for unique human experience cannot be a (fully) formalized or articulated process.

Use of lingual symbols

The use of symbols amplifies our intellectual abilities to a large extent, but by itself, it does not eventuate any new meanings.

> *"The mere manipulation of symbols does not in itself supply any new information, but is effective only because it assists the inarticulate mental powers exercised by reading off their result."* (Polanyi 1958, 83)

Namely, an explicit symbol does not mean anything by itself, according to the logic of affirmation; only a specific person can bestow it with meaning by using the symbol in a statement (ch. 3 and in more detail in ch. 8). The meaning of a symbol is determined by the previous tacit experiences of a given person. The use of symbols provides us—sometimes enormous—help in the reinterpretation and restructuring of these experiences.

Descriptive sciences such as zoology or botany do not rely on the manipulation of symbols but extend everyday speech with technical terms. (Polanyi 1958, 84) Polanyi highlights that in the case of these sciences, the condition for their development was rather the expansion of printed records. Exact and deductive sciences, however, lean increasingly on the second operational principle of language: increasing the measure of formalization and symbolic manipulation while decreasing the relevance of personal experience.

> *"Higher degrees of formalization make the statements of science more precise, its inferences more impersonal and correspondingly more 'reversible'; but every step towards this ideal is achieved by a progressive sacrifice of content."* (Polanyi 1958, 86)

This process takes us where the purely explicit and formal Laplacian ideal of knowledge does not correspond with the personal, tacit knowledge and experiences of man at all (ch. 1 and in more detail in ch. 6).

Tacit content

Polanyi uses the often quoted examples of riding a bicycle and medical diagnostics (e.g., understanding x-ray scans) to illustrate the sphere of ineffability—namely, that the scope of our unspoken knowledge always exceeds what our words are able to express.

> *"We may say in general that by acquiring a skill, whether muscular or intellectual, we achieve an understanding which we cannot put into words and which is continuous with the inarticulate faculties of animals."* (Polanyi 1958, 90)

Everything man can understand by obtaining an ability has existential meaning instead of denotative (referring to something). To be able to ride a bicycle is not a denotative declaration of something but an existential practice of knowledge. Animals seize every meaning this way due to their lack of articulated language; therefore, they are not able to formulate denotative statements. The rat perceives the maze in the same aspect. (Polanyi 1958, 90) (See ch. 4 and in more detail in ch. 8.)

Ineffable knowledge does not mean it cannot be discussed. A person may be able to give a detailed description of the movements he makes during cycling, but this description is no substitute for the knowledge gained in practice which he relies on when cycling. Therefore, we talk about *two different* types of knowledge: on the one hand, the existential meaning of the ability to ride a bicycle; on the other, the denotative knowledge, which is the communicable description of the ability to ride a bicycle organized into statements. (Polanyi 1958, 90)

Based on this, Polanyi makes a distinction between *hearing* and the *content* of a message. The former is *articulated* in a specific language. The latter, however, is completely *tacit*. Quite often, we cannot remember the articulated, explicit language and the factual formulation of a message, but we are still able to recall the tacit meaning of it. (Polanyi 1958, 91)

In general, we gain our knowledge of things through experience, just as animals do. Words gain their meaning through being articulated during experiences, such as when a mother tells her child, "This is a ball," while rolling the ball to him.

> *"Therefore, when I receive information by reading a letter and when I ponder the message of the letter, I am subsidiarily aware not only of its text, but also of all the past occasions by which I have come to understand the words of the text, and the whole range of this subsidiary awareness is presented focally in terms of the message."* (Polanyi 1958, 92)

Of course, we can learn new words from an encyclopedia or at school, but the meaning of these words does not actually come from the explicit words, but instead from our previous experiences that are jointly invoked upon listening to the defining words and focusing our understanding on the meaning of the new words we are learning.

> *"The conception in question is the focus of our attention, in terms of which we attend subsidiarily both to the text and to the objects indicated by the text."* (Polanyi 1958, 92)

We read a text instead of perceiving it. What we perceive is a line of ink twisting on a piece of paper. (Polanyi 1958, 92)

There can be a sharp difference between tacit understanding and operating with explicit symbols. In this case, we have to decide which we should rely on. The principles of Newtonian mechanics describe a reversible world in which, according to the Laplacian concept, any past or future condition is perfectly calculable from the actual description of the world and its mechanical principles. In this case, we ignore tacit content—namely that the world based on experience is irreversible—and stretch logical consistency to the extreme by applying the symbolic operations of Newtonian mechanics.

Whenever a cognitive scheme based on a new symbolic system comes into existence, it always brings systematic delusions along with it. Polanyi mentions animism, witchcraft, as well as oracles and taboo in connection with primitive oral cultures, which—as Polanyi approvingly notices—were replaced in Western culture by philosophy, theology, mathematics, and natural sciences, which led us to new systems of delusions such as the Laplacian ideal of knowledge, which Polanyi sharply criticizes. (Polanyi 1958, 93)

> *"The mind which entrusts itself to the operation of symbols acquires an intellectual tool of boundless power; but its use makes the mind liable to perils the range of which seems also unlimited."* (Polanyi 1958, 94)

The active principle controlling the intellectual endeavors of man

Polanyi declares that our intensive personal participation in the process of gaining and expanding our knowledge is an intellectual endeavor that shapes and justifies our perceptions and concepts—and as such, it comes from an *active principle*. This active principle moves and compels us to gain knowledge. It is present in the lowest animalistic levels, where we can discover its operation in sentience, alertness, as well as in perception. (Polanyi 1958, 96)

> *"This is the principle which guides all skills and connoisseurship, and informs all articulate knowing by way of the ubiquitous tacit coefficient on which spoken utterances must rely for their guidance and confirmation."* (Polanyi 1958, 132)

Polanyi, unfortunately, does not thoroughly explain what he means by the active principle controlling the intellectual endeavors of man, but, based on later discourses (especially in ch. 13), we can draw the conclusion that it is an emergent principle possessed by every single creature, and also an outcome of the ordering principles of emergence and evolution. It is important to highlight that it is not a vitalist but an *emergentist* concept, which, similarly to vitalism, denies the materialist concept of life.

BOX 12: Active, emergent principle

Of course, Polanyi does not use the expression "principle" in a positivist sense, where principles are only models and theories, serving the short and solid description of the phenomena of the material world. According to Polanyi, an active principle that controls the intellectual endeavors of man is an existing inherent aspect of reality. However, since Polanyi is an emergentist, this active principle is neither substantive nor the equivalent opposite of a substance, as it would be due to positivist or dualist interpretations, but an aspect or manifestation of reality that emerges in a specific material system. Polanyi, however, does not explain what he exactly means by it.

The relation of perception and language

Perception follows its self-defined standards and creates universal conceptual schemes. Such a conceptual scheme is, for example, that objects retain their sizes and shapes. (Polanyi 1958, 96-97)

> *"This tremendous generalization, on which we base our understanding of the universe, we have in common with the higher animals."* (Polanyi 1958, 97)

Polanyi sets up an analogy between the perceptive identification of things and the acquirement of words concerning them. We establish the *per se* imperceptible image of things according to different sensual impressions; in diverse situations, we shape the general meaning of words that can never be experienced in concrete situations. On top of that, lingual identifications are:

> *"[...] merely extend the theory of the universe implied in our sensory interpretations to the wider theory, implied in the vocabulary by which we talk about things."* (Polanyi 1958, 97)

Even a cat is able to recognize frogs. It creates a *tacit concept*, which forms the basis of how to think and act concerning the aspects of reality identified as frogs. Human articulation expands this perceptive ability we have in common with animals, thus opening wide the horizon of possibility to obtain knowledge concerning, e.g., frogs. Therefore, articulation is built on the tacit prearticulation foundation, which is present in both man and animals.

According to Polanyi's interpretation, both tacit ideas and articulated concepts exist. This perception, which we have in common with animals, is the source of our idiomorphic tacit ideas that form the basis of the ability—in both man and cat—to identify a particular living phenomenon as a frog. However, due to his articulated conceptual systems, man can massively deepen his tacit knowledge, which makes it possible, for example, to classify taxonomically and profoundly characterize different kinds of frogs, which is impossible for a cat.

While perception prefigures all our knowledge of things, drive satisfaction does the same with practical skills, and the two are always interwoven. In Polanyi's words:

> *"Though the information that we thus acquire—for example by eating, smoking, or making love—is necessarily centered on ourselves as agents, such information does in fact enter emphatically into our articulate picture of the world. To a disembodied intellect, entirely incapable of lust, pain or comfort, most of our vocabulary would be incomprehensible."* (Polanyi 1958, 99)

Drive satisfaction and perception are the primordial rudiments of two classes of intelligent behavior. The first (type A) manifests itself in practical, and the second (type B) in cognitive learning. Type C is formed by uniting the previous two. (Polanyi 1958, 100)

The rationality of our concepts

The obvious rationality of our concepts "is due to their being in contact with domains of reality, of which they have grasped one aspect." (Polanyi 1958, 104) Thus, our concepts are rooted in our tacit animal abilities and inherited tacit concepts, each of which grasps one aspect of reality. A cat is able to recognize a mouse. This process is not arbitrary, but instead a grasp of the existential meaning of a real, systematic scheme, justified by the fact that, while the cat acts according to its tacit knowledge, reality reconfirms the decisions accompanying this act in several and most unexpected ways. If the cat is smart enough, it will be able to gorge in the end.

> *"We grant authority over ourselves to the conceptions which we have accepted, because we acknowledge them as intimations—derived from the contact we make through them with reality—of an indefinite sequence of novel future occasions, which we may hope to master by developing these conceptions further, relying on our own judgment in its continued contact with reality."* (Polanyi 1958, 109)

Polanyi's evolutional concept of language

Polanyi makes a distinction between the reversible, routine-like use of language—which is assimilating experience into a fixed scheme of understanding—and the irreversible, heuristic use of it when in an opposite situation when language is adjusted to the new empirical challenges. Modifying our language also means modifying our intellectual spheres, since henceforth we construe our experience in this logically indescribable new scheme of interpretation.

> *"Indeed, any modification of an anticipatory framework, whether conceptual, perceptual or appetitive, is an irreversible heuristic act, which transforms our ways of thinking, seeing and appreciating in the hope of attuning our understanding, perception or sensuality more closely to what is true and right."* (Polanyi 1958, 106)

According to this, our language and the meaning of words change constantly, but since this change happens slowly and tacitly, we are hardly conscious of it. Similarly, different languages, developed by different groups of people in different historical ages while "groping in this world," are divergent and carry diverse explanations of the universe.

> *"In learning to speak, every child accepts a culture constructed on the premises of the traditional interpretation of the universe, rooted in the idiom of the group to which it was born, and every intellectual effort of the educated mind will be made within this frame of reference."* (Polanyi 1958, 112)

Therefore, different groups of people interpreting the circumambient world and its concerns, including their own experiences, on the grounds of a divergent set of words and grammar cannot understand the way other groups see and interpret our world and its details. Words are not simply products of convention.

> *"Our choice of language is a matter of truth or error, of right or wrong—of life or death."* (Polanyi 1958, 113)

For example, a cat that recognizes a dog as a mouse rather than as a dog could prove to be a perilous mistake.

Polanyi's criticism of Wittgenstein

Polanyi offers sharp criticism against the traditions of nominalist language perception that states that words are mere conventions, although open-structured. According to Polanyi, it would only make sense if they would also accept that the actual extensiveness of words is specified by a supervising principle that determines whether the words express the elements of reality that they are supposed to express. Namely, a word that could mean anything due to its open structure means nothing. However, such a supervising principle is not acceptable by nominalist traditions; thus, we cannot get an answer to the question: what is the determining factor in using a word for expressing an appropriate meaning? (Polanyi 1958, 113)

Polanyi also sharply criticizes Wittgenstein's concept of language-games, i.e., analysis of the use of language. According to him, it reveals nothing about the nature of things referred to by the language expressions. When Wittgenstein says, "'I don't know whether I am in pain or not' is not a significant proposition," it is actually a *deceptive substitution.* This is because the statement does not concern the nature of pain, its *existential* meaning, but the use of the word "pain," which only *denotes* pain. Thus, Wittgenstein's statement would be correct this way: "It is contrary to accepted usage to speak of pain whether I feel it or not." This is true but does not describe the nature of pain, that it includes—or not—the fact that sometimes I cannot tell if I am feeling pain or not. If Wittgenstein had put his statement in this latter way, he would have said something true that was irrelevant to the nature of pain, about which he was actually mistaken. (Polanyi 1958, 113-114)

> *"Correspondingly, disagreements on the nature of things cannot be expressed as disagreements about the existing use of words."* (Polanyi 1958, 114)

> *"I suggest that we should be more frank in facing our situation and acknowledge our own faculties for recognizing real entities, the designations of which form a rational vocabulary."* (Polanyi 1958, 114)

The classification of things over two millennia has not been done for our convenience but has resulted in serious evolutional advantages or disadvantages for a specific group of people.

Three strata of intentions

According to Polanyi's concept of language, *intention* (meaning) has three successive ever-deeper strata. (Polanyi 1958, 115-117)

1. The first comprises the readily specifiable properties that a class of things is known to share, apart from their common key feature. Unfortunately, Polanyi does not illustrate it with any examples.
2. The second stratum comprises the known but not readily specifiable properties that these things share. Polanyi illustrates it with the words "justice," "truth," and "courage." On the grounds of this relationship, he says:

> *"Understood in these terms, definition is a formalization of meaning which reduces its informal elements and partly replaces them by a formal operation (the reference to the definiens)."* (Polanyi 1958, 115)

   If we want to understand the meaning of justice, for example, it is good for us to define it, but the real existential meaning is found beyond the definition, which is only denotatively referred to by the word "justice." We can grasp the real meaning focally if we tacitly lean on past usages of the word in relevant situations.

3. The third and deepest level of intentions is formed by the indeterminate range of anticipations expressed by designating something.

Problem-solving

Analysis of problem-solving is important for Polanyi since, in the course of this process, he points out that animals are not simple mechanical structures but active and purpose-driven participants.

> *"We may say that the animal has seen a problem, if its perplexity lasts for some time and it is clearly trying to find a solution to the situation which puzzles it. In doing so, the animal is searching for a hidden aspect of the situation, the existence of which it surmises, and for the finding or achieving of which the manifest features of the situation serve it as tentative clues or instruments.*
> *To see a problem is a definite addition to knowledge, as much as it is to see a tree, or to see a mathematical proof—or a joke. It is a surmise which can be true or false, depending on whether the hidden possibilities of which it assumes the existence do actually exist or not."* K120

Problem-solving also has tacit roots since the realization of problems is a conjecture of a hidden rational aspect of reality, which, although nearly not as palpable as a tree, is just as real as it is.

> *"Nothing is a problem or discovery in itself; it can be a problem only if it puzzles and worries somebody, and a discovery only if it relieves somebody from the burden of a problem."* (Polanyi 1958, 122)

> *"The irreversible character of discovery suggests that no solution of a problem can be accredited as a discovery if it is achieved by a procedure following definite rules."* (Polanyi 1958, 123)

Accordingly, Polanyi separates the following stages of problem-solving:

1. *Appreciation;*
2. *Preparation;*
3. *Illumination;*
4. *Verification.*

From all of these, Polanyi draws the following conclusion:

> *"It follows that true discovery is not a strictly logical performance, and accordingly, we may describe the obstacle to be overcome in solving a*

> *problem as a 'logical gap', and speak of the width of the logical gap as the measure of the ingenuity required for solving the problem. 'Illumination' is then the leap by which the logical gap is crossed. It is the plunge by which we gain a foothold at another shore of reality. On such plunges the scientist has to stake bit by bit his entire professional life."* (Polanyi 1958, 123)

Polanyi's interpretation of the nature of problem-solving and discovery, along with their consequences, will be explained in the following chapter. Problem-solving has deep evolutional roots which originate from our most fundamental animal capabilities. When solving a problem, an animal executes an active heuristic act, which enables it to set foot in the previously unknown other side, thus performing an evolutional leap. This act cannot be interpreted as a passive mechanical one and cannot be compared to the simple process that evolves from the mechanism of natural selection (see ch. 13).

It is the active problem-solving that leads the animal towards genuine evolutional accomplishments and emergence. A detailed explanation of these, however, is found only in the final chapters (ch. 11-13).

> *"We have seen already that whenever we make (or believe we have made) contact with reality, we anticipate an indeterminate range of unexpected future confirmations of our knowledge derived from this contact. The interpretative framework of the educated mind is ever ready to meet somewhat novel experiences, and to deal with them in a somewhat novel manner. In this sense all life is endowed with originality and originality of a higher order is but a magnified form of a universal biological adaptivity. But genius makes contact with reality on an exceptionally wide range: seeing problems and reaching out to hidden possibilities for solving them, far beyond the anticipatory powers of current conceptions."* (Polanyi 1958, 124)

Conclusion

In this chapter, Polanyi's argument was to prove that man's intellectual capabilities have tacit roots. Articulation provides a conceptual and cogitative scheme whose operation is built entirely upon its own tacit foundations. It considerably extends and deepens these tacit intellectual foundations, which are practically the same as those of primates. Our articulated concepts are rooted in our most simple tacit capabilities, such as perception. Man has obtained a conceptual and intellectual tool in articulation that helps to significantly expand both the perception of reality and the knowledge related to it. Concepts and language grasp the aspects of reality, and every language carries a particular theory of the universe. This conceptual and cogitative scheme determines our perceptions and actions, rendering possible the creation of an even closer connection with reality to get hold of its deeper, inherent, rational correlations, which would be impossible to keep under our intellectual power purely by perception.

Chapter 6

Intellectual passions

Goals of the chapter

1. Polanyi shows that the tacit capabilities commonly shared by man, animal, and infants are multiplied in adults through articulation, which is, therefore, the foundation of human thinking that exceeds the intellectual life of animals.
2. Polanyi refers to higher-level human tacit abilities as intellectual passions and makes a distinction among three kinds: selective passion; heuristic passion (which he connects with creativity); and persuasive passion.
3. Leaning on the concept of intellectual passions, he analyzes science, technology, mathematics, art, and to a certain extent religion, in a uniform framework. An important statement is that in spite of the different purposes and content of certain areas, basically, it is the same intellectual passion that enables us to operate in them.
4. He points out the connection between intellectual beauty and truth. According to him, the intellectual beauty of a scientific theory is a sign of its deep relationship with reality; that is, the sign of truth.
5. He explains scientific breakthroughs and technological inventions with the dynamics of intellectual dwelling in and breaking out: heuristic passion transforms scientific or technological knowledge by a logically unspecifiable leap.
6. According to Polanyi, in spite of its every formal excellence, science is directed by tacit intellectual capabilities; thus, in contrast with positivist concepts, science cannot present any unusual perfection since it is not a completely exact, formalized logical knowledge machine, but an important subsystem of human culture, just as arts, religion, law, or morality.
7. Therefore, in this aspect, science is not special at all; however, in the Western world, science still holds the privilege to set out the direction to future thought concerning our whole culture. Polanyi prefigures that, as a consequence, similarly to Greek culture, Western culture might as well disappear if it is unable to recognize its proper foundations, values, and alignments.

***

"Science can then no longer hope to survive on an island of positive facts, around which the rest of man's intellectual heritage sinks to the status of subjective emotionalism. It must claim that certain emotions are right; and if it can make good such a claim, it will not only save itself but sustain by its example the whole system of cultural life of which it forms part." (Polanyi 1958, 134)

Scientific intellectual passions

First of all, Polanyi clearly declares that:

"I want to show that scientific passions are no mere psychological by-product, but have a logical function which contributes an indispensable element to science." (Polanyi 1958, 134)

According to this, scientific intellectual passions have three fundamental logical functions:

1. *Selective function* tells which of the scientific facts and definitions are relevant or irrelevant. From the whole of possible facts, only certain facts are relevant. The number of these is dwarfed by the literally uncountable occurrences of the complete number of facts.
2. *Heuristic function* connects a vision of reality with scientific facts, thus creating a new scientific discovery—which can be a statement but also a new conceptual system.
3. *Persuasive function* motivates the discoverer to share the new discovery with the rest of the scientific community, and also to make them accept it.

Selective function

At first, Polanyi describes the selective function of intellectual passions, which points out what is considered important from a scientific point of view.

"The excitement of the scientist making a discovery is an intellectual passion, telling that something is intellectually precious and, more particularly, that it is precious to science." (Polanyi 1958, 134)

Every discoverer can feel the intellectual thrill and joy accompanying these new revelations. Less significant recognitions meet cold reception at best, while more relevant ones throw new light upon things, and their acknowledgment means that there is no way back to former theories. In the objectivist concept of science—which Polanyi also criticizes—intellectual passions are just psychological by-products with no real role in the process of scientific discovery. Polanyi, however, claims that without the guidance of intellectual passions, leaning exclusively on logic in a formal method, we would never be able to distinguish scientifically relevant from irrelevant details. If we were to put aside our tacit capabilities, such as intellectual passions, any explicitly definable event would appear to us as a scientific fact with no difference in importance. It is an obvious critique of the Laplacian vision, which tries to perceive our world on a purely objective basis—first presented by Polanyi on the first page of *Personal Knowledge* (see ch. 1) and now revisited. Upon applying the Laplacian principle, there would be neither enough time nor resource to research mundane life or humanity since every tiny slice of the universe should be investigated with the same thoroughness.

> *"Scientific passion serves also as a guide in the assessment of what is of higher and what of lesser interest; what is great in science, and what relatively slight."* (Polanyi 1958, 135)

Of course, the question is: what is the deciding factor for intellectual passions when choosing the facts of real value and meaning from the vast ocean of scientifically irrelevant and meaningless facts? Polanyi's answer is as follows:

> *"Our vision of the general nature of things is our guide for the interpretation of all future experience."* (Polanyi 1958, 135)

This tacit vision of reality—which is not yet knowledge, only the possibility for it—directs our scientific sense of beauty and intellectual passions.

> *"Our vision of reality, to which our sense of scientific beauty responds, must suggest to us the kind of questions that it should be reasonable and interesting to explore."* (Polanyi 1958, 135)

Humanity, and, on a lower level, animals, also possess the ability to recognize and solve various problems that express a portion of rational correlations of reality with either lower or higher relevance. On a higher intellectual level, this ability of *problem recognition* starts to operate by the contribution of our scientific sense of beauty: led by the prearticulation tacit vision of reality, we find our way through our intellectual passions towards the discovery of scientific answers. In the event that these answers are true, they prove beautiful, full of meaning, and valuable in a scientific sense. (Polanyi 1958, 135)

Heuristic function

The next objective for our intellectual passions in science is to fulfill their heuristic function. Polanyi correlates this with synthesis and creation: on the grounds of heuristic function, the recognition of scientific value "melts" with the ability to discover scientific value, thus creating something entirely new. (Polanyi 1958, 143)

Scientific discovery is an irreversible, passionate, creative process that fundamentally changes our scheme of interpretation. Thus, the new scheme of interpretation cannot originate from the old since the two are not related in terms of deduction. (Polanyi 1958, 143)

Polanyi illustrates the passionate uniqueness of discovery with repeated examples of Einstein's and Kepler's overjoyed states of mind. Regarding Kepler, he makes a distinction between scientific assumptions proven wrong and unscientific conjectures considered incompetent. Kepler, who discovered the elliptical shape of planetary orbits, is a great subject to present fine examples for both. As an illustration of the latter, Kepler considered the sun to be a sentient intellect listening to the music of spheres, and he made a musical notation of the musical patterns expressed by the proportions of planetary orbits (see ch. 1). (Polanyi 1958, 143-144)

Polanyi is convinced that intellectual beauty is the *sign of reality*. (Polanyi 1958, 145) Our preliminary vision of reality and intellectual passions only leads us towards subsequent discoveries during scientific research if they are not subjective or random but point towards the development of a closer relationship with reality.

> *"Truth lies in the achievement of a contact with reality—a contact destined to reveal itself further by an indefinite range of yet unforeseen consequences."* (Polanyi 1958, 147)

Our intellectual sense of beauty is the manifestation of the tacit animal capabilities in articulated scientific conceptual and explanatory systems. Thus, the fundamental animal and human abilities to recognize and grasp reality manifest in articulated systems to become tacit capabilities such as the ability to recognize intellectual beauty. Most scientists accepted the theory of relativity because of its intellectual beauty (see also ch. 1); empirical observations only later started to support its reality.

Regarding this, Polanyi sharply criticizes the positivist concepts that put reality and the connection with it into brackets due to its indefinable, not wholly explicable, not formalizable nature, claiming that scientific truth manifests only in the fruitfulness of scientific ideas. Polanyi's answer to this is that the Copernican system was more than fruitful, but it came to be considered as fruitful because of its reality and because it carried several other unforeseen assumptions of reality as a vision of a new reality.

To say that the Copernican system was fruitful means nothing in itself, though; the statement gains its real meaning if we tacitly add that the system was fruitful from the aspect of reality. This is the only way we are able to distinguish the fruitfulness of the Copernican system from the fruitfulness of astrology (as a source of income) or Marxism (as a source of authority). If we do not declare, but just tacitly mean, that the Copernican system was fruitful because of its truth, we also perform a deceptive substitution, a "Laplacian sleight of hand." (Polanyi 1958, 146-148)

> *"By the time Newton published his Principia anyone could see that Copernicus had been right; but Copernicus and the Copernicans, and among these Newton himself, were convinced of it long before this. The veridical quality of the system, of which they were then convinced, could not be its subsequently observed fruitfulness. [...] The mark of true discovery is not its fruitfulness but the intimation of its fruitfulness."* (Polanyi 1958, 148)

On the grounds of all of this, Polanyi arrives at the following conclusion: 1. intellectual beauty reveals the truth about nature; 2. this beauty cannot be identified with any (explicit) attractivity; 3. it is very difficult to distinguish the two. (Polanyi 1958, 149)

Persuasive function

An additional aspect of our intellectual passions in science is the persuasive function, which arises from the following:

> *"[...] we suffer when a vision of reality to which we have committed ourselves is contemptuously ignored by others."* (Polanyi 1958, 150)

Heuristic discovery leads us to a vision of reality that is separated from other perceptions of reality by a *logical gap.* (Polanyi 1958, 151) The objective of the persuasive function is to build a bridge over this gap. The intellectual passion of persuasion compels us to make others know and accept our discovery, to which we are already committed. (Polanyi 1958, 150) This is quite a difficult mission, however, because:

> *"Formal operations relying on one framework of interpretation cannot demonstrate a proposition to persons who rely on another framework."* (Polanyi 1958, 151)

Another obstacle in the way of persuasion presents itself when another researcher, committed to a different conceptual system, does not make any effort to open himself to the arguments of our own conceptual system. To at least consider the weight of our argument, his first efforts should be to pick up the conceptual system and language of this new interpretational scheme that is, actually, meaningless for him.

Furthermore, man cannot learn a new language without trusting in it first. But researchers, committed to contradicting conceptual systems are typically distrustful towards each other. Thus, they restrain from accepting another conceptual and interpretational system, fearing that if they do so, they may feel an obligation to accommodate, at first, those bizarre-looking arguments, and then a whole repulsive conceptual system.

> *"Proponents of a new system can convince their audience only by first winning their intellectual sympathy for a doctrine they have not yet grasped. Those who listen sympathetically will discover for themselves what they would otherwise never have understood. Such an acceptance is a heuristic process, a self-modifying act, and to this extent a conversion. It produces disciples forming a school, the members of which are separated for the time being by a logical gap from those outside it."* (Polanyi 1958, 151)

Therefore, persuasion is not an easy task, and it is impossible to argue successfully while being on the grounds of the other person's conceptual system. Persuasion has to start with raising intellectual sympathy, which can be achieved by presenting intellectual beauty during the persuasive process. Thus, persuasion has to be implemented by forms that lead to the heuristic understanding and later the acceptance of the suggested conceptual system. (Polanyi 1958, 151) It is, therefore, not only the objective, impersonal comparison of facts since the arguing parties are committed to their own standpoints and thus to their own conceptual system. Such situations inevitably lead to the offense of the other person, which Polanyi considers as regrettable but sometimes inevitable. (Polanyi 1958, 151-152)

BOX 13: Thomas Kuhn and Polanyi

It is not unusual to interpret Polanyi's scientific views in the framework of Kuhn's more well-known, thus more effective, paradigm theory, or at least to assume that there is a close similarity between the two concepts. This similarity is, however, just superficial. Both authors examine the history of science, trying to conceive how science actually works. Both declare that the paradigms/conceptual systems that develop on the explicit level are not comparable due to serious methodological, conceptual, and institutional differences. However, one of the substantial points of Polanyi's theory is that the development of science is characterized by the tacit foundations and especially by the well-developed intellectual passions that are common in every scientific paradigm beyond the incomparable explicit level. According to Polanyi, there is perpetuity in the development of science because of this; however, in contrast with positivist ideas, it is neither exclusively theoretical nor limited to the constantly growing number of objective facts. Kuhn's theory, on the other hand, does not have any tacit fundaments similar to Polanyi's.

Polanyi mentions four examples of debates concerning fundamental convictions among scientists committed to contradicting conceptual systems around the history of Western science. The first is the often mentioned collision of Ptolemaic and Copernican systems, where the most important figure was Galileo Galilei. In Polanyi's interpretation, "The real ground of Galileo's conviction lay in his passionate appreciation of the greater scientific value of the heliocentric view," (Polanyi 1958, 152) and not in the superiority of his arguments, nor in the higher number of objective facts, he could recite in support of the Copernican system. Counter to that, there was the plausible but later falsified experience of the earth resting still under our feet, and the moral conviction—which Polanyi also accepts—to "retain for man a location which corresponds to his importance in the universe," (Polanyi 1958, 152) which was already compromised by the early formulation of the Laplacian ideal of knowledge. The victory of Copernicanism then "established the principle that scientific truth shall take no account of its religious or moral repercussions." (Polanyi 1958, 153) However, Polanyi adds the following:

> *"My opposition to a universal mechanical interpretation of things, on the ground that it impairs man's moral consciousness, also implies some*

> *measure of dissent from the absolute moral neutrality of science."* (Polanyi 1958, 153)

This does not mean, however, that scientific truth has to be cast aside to avoid negative moral consequences, since in that case, our starting point would be the theory of dissevering science from morality, which says that, irrespective of moral justice, it is possible to reach scientific truths that contradict the fundamental laws of morality.

However, Polanyi states that morality and science are not inseparable because, in both domains, we are led by personal tacit passions which, in the domain of science, appear in our thinking, decisions, and actions as intellectual passions and, in the domain of morality, as moral passions (see in detail in ch. 7). Consequently, owing to the fact that collective tacit foundations were present even in the early forms of evolution, scientific and moral truths are in concordance with each other thus, proper scientific ideas do not contradict proper moral commitments. The Laplacian ideal of knowledge, however, not only disagrees with certain moral truths but, beyond that question, even the reality of human morality, which is, according to Polanyi, a sign of it not being true.

Therefore, Polanyi does not suggest tossing aside scientific truths just because they disagree with certain truths of human morality. Based on his conviction, if we were to put our concept of science to proper foundations—namely, if we were to acknowledge the idea of personal knowledge—and as a part of that, we also accepted the scientific role of intellectual passions, most of the scientific discoveries whose consequences contradict morality would prove wrong, thus could not be considered as scientific truths. The number and significance of such conflicts could decrease by a great measure if science could admittedly operate on tacit foundations. (Polanyi 1958, 153)

Empirism, according to Polanyi, is a result of the victory of the Copernican concept and the works of Galilei. Polanyi also does not agree with it without any consent as its consistent following would make the operation of science nonsensical, similar to the Laplacian ideal of knowledge. However, he acknowledges that empiricism has played an important role in the development of science, and, when considering it as an effort to understand material conditions thoroughly, it becomes an important part of personal knowledge, and thus it is easily acceptable for him. (Polanyi 1958, 153)

The other three examples that Polanyi brings from the history of science are the case of the so-called Bode's Law, the mixed reception of van 't Hoff's doctoral thesis, and a controversy about the nature of alcoholic fermentation from the nineteenth century, which interwove with the doctrine of spontaneous creation of life.

Bode's Law expresses a mathematically describable rational correlation of nature, converting the distance of the planets from the sun into a sequence. By accepting the law, it can be claimed that there must be a planet between Mars and Jupiter. In contrast, the young Hegel set up a randomized correlation that correctly reflected the distances of known planets from the sun without denoting the discovery of a new planet. Hegel claimed that hypothesizing the existence of a new planet using any fantastic mathematical interrelation is unscholarly mirage-chasing. However, the astronomers soon discovered Ceres and then many other small planets in the region marked by Bode's Law. In Polanyi's consideration, Bode's Law may not have any rational foundations—in the fifties, it was not possible to clearly determine this—but it surely was a professional, scientific assumption that later proved true. However, Hegel's claim was altogether unscientific and incompetent since it rejected any rational coherency of nature concerning the location of planets. (Polanyi 1958, 153-155)

BOX 14: Bode's Law or Titius–Bode Law

The planets of our solar system move along fixed orbits, at a specified distance from the sun and from each other. In the first half of the eighteenth century, it was already evident that the distance of planets from the sun follows some kind of sequence, which was ultimately mathematically formulated by Johann Elbert Bode in 1778 and, according to this, one planet was missing. The quest to find Ceres and the rest of the small planets orbiting between Mars and Jupiter started at the beginning of the nineteenth century, and soon they were found. Bode's Law is not a rule of nature; there is no particular correlation that it can be linked with. However, according to the models describing the formation of the solar system, one could presume that the forming of planets had a serious effect on the remainder coming into existence—for example, one with the size of Jupiter "cleaned" space around itself, preventing other planets from coming into existence, and due to its significant mass, it strongly influenced the orbit of the rest of the planets—which means that mutual distance of the developed orbits could hardly be randomized. With significant numbers of exoplanets being discovered, we can assume that a rule similar to Bode's Lawiels is applicable in further solar systems too. According to this, contemporary astronomers talk about Bode's Generalized Law, which is not applicable only to our solar system.

Van 't Hoff's doctoral thesis suffered sharp attacks because it was expressly speculative, and at the time, it was conceived that only a small number of scientific facts seemed to support it. Regarding this, Polanyi notes:

> *"In spite of the fact that chemistry is largely based on the speculations by Dalton, Kekulé and van 't Hoff, which were initially unaccompanied by any experimental observations, chemists still remain suspicious of this kind of work. Since they do not sufficiently trust themselves to distinguish true theoretical discoveries from empty speculations, they feel compelled to act on a presumption which may one day cause the rejection of a theoretical paper of supreme importance in favor of comparatively trivial experimental studies."* (Polanyi 1958, 156)

Polanyi acknowledges that it is quite difficult to distinguish the two purely by the criteria of empiricism. (Polanyi 1958, 156)

BOX 15: Jacobus Henricus van 't Hoff

van 't Hoff was a Dutch chemist, the first person to receive the Nobel Prize in Chemistry in 1901. Van 't Hoff's doctoral thesis mentioned by Polanyi formulated the speculative hypothesis in 1874 that the chemical bonds between carbon atoms and their neighbors were directed towards the corners of a regular tetrahedron. After 1880, this conception became the basis of every experiment seeking the explanation of the three-dimensional structure (stereochemistry) of organic compounds. However, in the 1870s, there were sharp attacks from the part of the scientific community against both the hypothesis and van 't Hoff himself, stating that because of the lack of any empirical experiment, his hypothesis has no scientific foundation.

Polanyi draws the following conclusion based on the described events of the history of science: the idea of the moral neutrality of science, the idea of empiricism, and the appearance of determinative contemporary scientific ideas are all derived from debates of the past. Thus, they are not carved in stone, and their iterative reinterpretation cannot be avoided, which can lead to their occasional rejection. Reinterpretation has to happen on the basis of our personal evaluation since the purpose of reinterpretation is always to shape our intellectual life.

Definition of scientific value

Scientific value is the joint outcome of three contributing factors. (Polanyi 1958, 135-136) These are:

1. *Certainty (accuracy);*
2. *Systematic relevance (profundity);*
3. *Intrinsic curiosity.*

Polanyi uses the example of neo-Darwinism to illustrate his train of thought. According to his evaluation, neo-Darwinism is perfectly fit for points 2 and 3 since it relates to a subject—the origin of man—of the utmost intrinsic interest, and in addition to that, it beautifully fits into the mechanistic image of the universe conceived by science. Although its certainty is quite low—Polanyi grades it as insufficient—in terms of its strong features in points 2 and 3, it qualifies for acknowledgment as a valuable scientific idea. (Polanyi 1958, 136)

Though not definable in precise terms, the scientific value can, as a rule, be reliably assessed. (Polanyi 1958, 136) Think about the inner interestingness of things or phenomena. A living animal is more interesting to us than a dead one, a cat is more interesting than a mosquito, and a man is more interesting than a cat. (Polanyi 1958, 139) This evaluation is obvious as it is rooted in our pre-scientific tacit sensitivity. Researching exact explicit criteria would lead us to think—as so many others do on the grounds of objectivism and positivism—that man cannot be more interesting than a mosquito. On a similar basis, we could decide that man's place in the universe or his moral decisions are neither interesting nor important at all, compared to such "exciting" questions as the operation of the human digestive system or the regulators of overnight hormone production.

The Laplacian ideal of knowledge

The pursuit of objectivity within science takes us to the Laplacian ideal of knowledge, which we have already mentioned several times (especially in chapter 1, BOX 1). This ideal of knowledge tries to accomplish the absolute impersonality of science (objectivity) "by representing the world in terms of its exactly determined particulars." (Polanyi 1958, 139) However, Polanyi's standpoint is:

> *"This ideal of universal knowledge is mistaken, since it substitutes for the subjects in which we are interested a set of data which tell us nothing that we want to know."* (Polanyi 1958, 140)

First of all, we are interested in living beings, men, and human decisions, not in the complete and perfectly explicit atomic description of the universe. The Laplacian ideal of knowledge, however, can be attractive for us because it substitutes the knowledge of all experience for the knowledge of all atomic data through deceptive substitution (see ch. 1). (Polanyi 1958, 140-141) The only way to gain any experience or knowledge about comprehensive (emergent) objects like cats, frogs, buildings, machinery, etc., is by our preliminary tacit knowledge. The highly theoretical knowledge of explicit details—which is not connected to concrete perceptions—is different from the comprehensive experimental impressions or knowledge resulting from the active tacit endeavors of animal or man. We are only able to say anything interesting about explicit details if we have a preliminary idea about the comprehensive whole to which these details are attached. However, we cannot get hold of such comprehensive ideas by sticking to the Laplacian ideal of knowledge, since these are typically pre-scientific—a fact to which Polanyi draws our attention with a pinch of irony: "The existence of animals was not discovered by zoologists, nor that of plants by botanists." (Polanyi 1958, 139)

Thus, this deceptive substitution makes the Laplacian ideal of knowledge usable, but this same step also hides the truth—which is quite inconvenient for modern critical science—that knowledge is personal and tacit, and scientific knowledge ensues from our pre-scientific tacit sensitivity about things.

Polanyi considers the objectivist approach highly dangerous because it denies the role of intellectual passions—which, in reality, are responsible for the operation of science. If consistently applied, this conception could even lead to the discrediting of science, as already happened, albeit for a different reason, in the times of St. Augustine. (Polanyi 1958, 141) However, according to Polanyi, the real danger of the Laplacian ideal of knowledge does not reside in this but in its effect on our image of man.

> *"Its reductive program, applied to politics, entails the idea that political action is necessarily shaped by force, motivated by greed and fear, with morality used as a screen to delude the victims. […]*
> *Applied to human affairs, the Laplacian universal mechanics induces the teaching that material welfare and the establishment of an unlimited power for imposing the conditions of material welfare are the supreme good."* (Polanyi 1958, 141-142)

All of this is driven by the passion for perfectly objective, impersonal knowledge. However, the operation of this passion can never be exposed, as it can never be in accordance with the ideal of knowledge. That is why it is necessary to formulate the purpose of science along with aspects of utility with deceptive substitutions.

> *"A discovery will then no longer be valued by the satisfaction which it gives to the intellectual passions of scientists, but will be assessed according to its probable utility for strengthening public power and improving the standard of living. Scientific value will be discredited and its appreciation suppressed."* (Polanyi 1958, 141)

Our image of science, however, has to give an account of the actual operation of science, namely, about our tacit intellectual capabilities and passions that actually keep science in operation.

The premises of discovery

The premises of discovery are certain tacit sentiments and purposes that are only present implicitly before the discovery—namely, before the leap of discovery—over the logical gap. They can only be explicated and articulated after the discovery. (Polanyi 1958, 161)

Natural sciences mostly deal with facts that originate from our common pre-scientific knowledge, for example, the fact that frogs evolve from tadpoles. Therefore, factuality by itself is not yet a science, since a significant number of facts are pre-scientific and, on the other hand, the unscientific conceptual systems like astrology or magic also deal with facts, although science denies their vast majority as it does not consider them as scientific facts.

Only specific facts can be considered as scientific, but—as we have seen—a vast majority of these ensue from our pre-scientific tacit knowledge. (Polanyi 1958, 161) Thus, we are not able to know the scientific, logical premises of factuality before starting to determine facts while orienting ourselves in the world. On the contrary, we realize the premises of factuality while reflecting on ways of identifying such facts that can stand the trial of science. In the beginning, there are facts whose existence we believe in—for example, that frogs evolve from tadpoles—followed by the exploration of the explicit premises that can be used to conceive the practical principles of fact-finding. On top of that, the evaluation of scientific facts is not irrespective of the conceptual systems constituting the basis of their evaluation, "for within two different conceptual frameworks the same range of experience takes the shape of different facts and different evidence." (Polanyi 1958, 167)

Polanyi revisits the examples of swimming and riding a bicycle (see ch. 4) to illustrate that the premises (maximas) of a skill cannot be focally observed and explicitly described before practicing the skill, as they are not present unless the skill is learned and practiced. A result of that is that the observation of premises is only possible if, with the help of these premises, we have already tacitly learned and practiced the skill whose premises we would like to know focally. (Polanyi 1958, 162) Therefore, it is no paradox at all that we are only able to focally get hold of the premises of the discovery after the actual discovery. Evaluation of science is especially difficult because no one can see through the whole of the knowledge embedded in science. A theoretical physicist dealing with string theory has no connection to an epidemiologist researching contagious tropical diseases. Most probably, he has no real idea about the most basic facts and principles of this other field and vice versa. In cases like this, we necessarily lean on the authorities of common scientific life, accepting their proficiency in the given field of profession. This is how those relationships interlace the entirety of science, creating links to the past, too, since a part of scientific authorities on whose opinion we lean are deceased. Consequently, science makes up an extensive *tradition.*

> *"Its members recognize the same set of persons as their masters and derive from this allegiance a common tradition, of which each carries on a particular strand."* (Polanyi 1958, 163)

Polanyi is committed to the understanding of science as a comprehensive whole. Acknowledgment of authority, however, does not mean unconditional comprehensive acceptance of every detail. Nevertheless, those who would cast aside the constitution of science as it is do not deserve to be called scientists. However, authority as an institution does not stop the constant reformation of science since it is one of the fundamental attributes of science. A precondition to this, with regard to certain—even fundamentally important—details, is that the reigning authority is challenged and defeated in a debate. Polanyi does this, for example, in the case of the comprehension of truth, by explaining his philosophy of science. (Polanyi 1958, 163-164)

According to Polanyi, however:

> *"To acknowledge a person as a scientist—and even as a very great scientist—is merely to acknowledge him as competent in science, which admits the possibility that he was, or is, in many ways mistaken."* (Polanyi 1958, 164)

> *"In this sense then science embraces a consistent pursuit of gradually changing, and—I believe—on the whole, evermore enlightened and elevated intellectual aspirations."* (Polanyi 1958, 164)

The false objectivist ideal of science inclines scientists to give completely inadequate, explicit formulations concerning the idea of science and its fundamental principles, like "limit the term 'science' to propositions that we regard as valid, and the premises of science to what we consider to be its true premises." (Polanyi 1958, 164). These explicit formulations, however, obtain their meaning during tacit affirmations (see ch. 8). According to this:

> *"A scientist can accept, therefore, the most inadequate and misleading formulation of his own scientific principles without ever realizing what is being said, because he automatically supplements it by his tacit knowledge of what science really is, and thus makes the formulation ring true."* (Polanyi 1958, 169)

According to Polanyi's understanding, science does not develop due to the impersonal application of certain exact formulas, but, rather, is an enormous *system of convictions,* the roots of which go back to the distant past; hence, it cannot be illustrated by the use of explicit, objective ideas. (Polanyi 1958, 170-171)

Animal and intellectual passions

Polanyi compares animal and intellectual passions. According to his standpoint, animal desires cease upon completion; for example, hunger ceases upon eating. However, intellectual passions uphold themselves by their own fulfillment. When a scientist captures a correlation of nature, he will always admire its beauty, just as a great stage performance provides constantly repeated intellectual pleasure to those who understand it. These cases exemplify that "the articulate powers of man can extend the range of such joys to whole systems of cultural gratification," (Polanyi 1958, 173) since "these passions are public, not private: they delight in cherishing something external to us, for its own sake." (Polanyi 1958, 174) Thus, animal passions are *instinctive* and *private,* while intellectual passions are of a *cultural* nature in terms of their orientation; e.g., they pursue scientific discoveries that, due to one's curiosity or relevance, can bring intellectual gratification to others too. Intellectual passions are always of a *public nature.*

In contrast to the dualist perception of intellect, which considers the mind to be an individual rational force independent of bodily passions, Polanyi admits that similarly to our bodily animal life, our intellectual life is also directed by passions attached to our body. Contrary to materialist monism, he affirms that "while appetites are guided by standards of private satisfaction, a passion for mental excellence believes itself to be fulfilling universal obligations." (Polanyi 1958, 174) This actually means that Polanyi does not consider the intellectual passions rooted in the body and defines the intellectual life of man to be of a material nature. To the question, "Why do we have to accept this distinction?" he gives a morally motivated answer, which he explains later in ch. 7.

> *"This distinction is vital to the existence of culture. If it is repudiated, all cultural life becomes subordinated in principle to the demand of our appetites and of the public authorities responsible for the advancement of material welfare."* (Polanyi 1958, 174)

Discovery and invention

Polanyi defines pure or self-supporting knowledge and technology as opposites. In this comparison, technology is the knowledge that aims to create practical change, which means (bearing an existential meaning; see ch. 4) expedient performance. In contrast, pure knowledge does not seek to bring any practical change, but instead, it denotes certain things as true or false (denotative meaning; see ch. 4). (Polanyi 1958, 177)

> *"Technology teaches only actions to be undertaken for material advantages by the use of implements according to (more or less) specifiable rules. Such a rule is an operational principle. As implements are defined and understood in terms of an action which they serve, they are defined and understood likewise in terms of the operational principle which tells how to perform such an action."* (Polanyi 1958, 176)

BOX 16: Polanyi and the Patent Reform Act

In the '30s and '40s, Polanyi paid a lot of attention—not only from a philosophical point of view—to the issue of patents. He even conceived a specific proposal about how to reform patent law for the sake of economic efficiency (*Patent Reform: A Plan for Encouraging the Application of Inventions*). He dealt with such topics as the question of inheritability of patent rights, or the ownership rights of a patent when the intellectual product was conceived by an employee who utilized his company's resources to create that product. Polanyi's suggestions were received by the professional medium with a considerable lack of interest.

Patent law makes a distinction between *discovery* and *invention*. While discovery involves something that already exists but was previously unknown and increases our knowledge of natural law, invention involves creating something totally new with one's own ideas and development and establishes a new operational principle. Similar to discovery, the invention also involves a leap through a logical gap that lies between the already known operational principles and those that are just on the brink of discovery. (Polanyi 1958, 177) A technological invention, however, has to be economical, though, in pure science, this is not a requirement; for example, inventing a new type of steam engine in today's economic and social climate is not possible due to its lack of economic value. (Polanyi 1958, 177)

Thus, we discuss here the two subtypes of problem-solving:

> *"The heuristic passion of the technician centers therefore on his own distinctive focus. He follows the intimations, not of a natural order, but of a possibility for making things work in a new way for an acceptable purpose, and cheaply enough to show a profit."* (Polanyi 1958, 178)

> *"The scientific method was devised precisely for the purpose of elucidating the nature of things under more carefully controlled conditions and by more rigorous criteria than are present in the situations created by practical problems."* (Polanyi 1958, 181)

Explication of the rules of professions practiced in guilds, namely the destructive analysis of sub-processes, led to the emergence of the modern factory industry. (Polanyi 1958, 175) This created the opportunity to interconnect the two. (Polanyi 1958, 182)

Since the development of the factory industry, science has played a significant role in configuring techniques, and parallel to that, techniques based on applied science created their own scientific system, which probably would not have been able to develop within the framework of pure science. Polanyi distinguishes two subtypes within the border spheres of science and technology:

1. *Systematic technologies* which can be cultivated in the same way as pure science (e.g., electronics, aerodynamics);
2. *Technically justified science*, which mostly means the scientific examination of materials whose role is considered industrially important (e.g., research of oil, carbon, cotton, etc.).

Regarding the Soviet example, Polanyi highlights that the materialist and profit-oriented perception of science questions the demarcation of science and technology. In a paradoxical way, this perception of science ensues from excessive moral aspirations, as Polanyi discusses in detail in ch. 7. Remaining with the Soviet example, he notes that materialist perception of science can be generalized to the whole of the culture: "[...] materialistic interpretation of culture is a disguised imperative: it both declares that culture really is, and decrees that it ought to be, the servant of welfare." (Polanyi 1958, 180) Likewise, the pursuit of science for its own sake is considered immoral since science is obviously practiced for the purpose of financial gain in the Soviet Union. (Polanyi 1958, 180-181)

Utilitarian perception of science is a hazard to the pursuit of science, which is not necessarily obvious at first sight because the presence of science in the Western world of the twentieth century seems to be natural. However, this situation is not as old as we would think, and it heavily leans on the social environment surrounding it as a condition of opportunity. Radical changes such as the transformation of the Soviet Union deem the pursuit of science impossible (Polanyi's detailed analysis of the negative consequences of Marxism follows in chapter 7). (Polanyi 1958, 181) Offensiveness towards science accompanies its complete history anyway.

> *"Encircled today between the crude utilitarianism of the philistine and the ideological utilitarianism of the modern revolutionary movement, the love of pure science may falter and die."* (Polanyi 1958, 182)

> *"No important discovery can be made in science by anyone who does not believe that science is important—indeed supremely important—in itself."* (Polanyi 1958, 183)

Mathematics

Mathematics is the extension of perception that creates an area of research that is irrespective of our direct experiences and interesting for its own sake. (Polanyi 1958, 186)

According to Polanyi, the general perception that mathematics is only a set of formal tautologies is false since tautologies are necessarily true, but mathematics is not. (Polanyi 1958, 187) A mathematician is driven by intellectual beauty to accept mathematics as true, in spite of the fact that mathematics is not a logical necessity. (Polanyi 1958, 188) Intellectual beauty, however, is actually a sign of hidden reality. In mathematics, the correctly captured reality does not appear in terms of sensory experience but instead, unexpectedly, in the branches of mathematics, irrespective of any experience. (Polanyi 1958, 189-190)

Acknowledgment of mathematical premises and axioms is not an arbitrary act but instead is based on the tacit, adjunct understanding of theorems. (Polanyi 1958, 191) Declaring that mathematical statements are strictly proven and that mathematics is worth dealing with since it is a perfectly exact science is the dismissal of mathematics since it is built upon such axioms and premises which are not proven in the strictest sense of the word. (Polanyi 1958, 191-192)

> *"We should declare instead candidly that we dwell on mathematics and affirm its statements for the sake of its intellectual beauty, which betokens the reality of its conceptions and the truth of its assertions."* (Polanyi 1958, 192)

Dwelling in and breaking out

Polanyi starts the last subchapter with the following declaration:

> *"A valid articulate framework may be a theory, or a mathematical discovery, or a symphony. Whichever it is, it will be used by dwelling in it, and this indwelling can be consciously experienced."* (Polanyi 1958, 195)

Here Polanyi talks about the assimilation of our intellectual instruments—e.g., when we make a hammer a part of our body by using it as an extension to our body after acquiring the art of hammering. We do the same with our diverse concepts and articulated schemes since we are living our spiritual life within the frames of interpretation held up by them. Polanyi refers to this as *dwelling in.*

Polanyi does not make a substantive distinction between the operation of science and other articulated systems. Artistic beauty is the sign of artistic reality, just as mathematical or scientific beauty is a sign of mathematical or scientific reality. (Polanyi 1958, 201) All of these are driven by tacit intellectual passions, and all include the acknowledgment of the intellectual framework within which our cognitive processes happen. The human mind, however, is essentially apprehensive; thus, it does not settle for this dwelling in and attempts to extend the articulated schemes or, in a certain sense, to penetrate and recreate them logically. Polanyi refers to this as breaking out, which he describes as a temporal but intensive spiritual experience in the course of which we have the opportunity to quasi contemplate our former spiritual dwelling place from the outside, while upon the heuristic transformation of it, we return to a new spiritual dwelling place. Polanyi's best and most thoroughly explained example, of course, concerns scientific discovery. (Polanyi 1958, 195-196)

We interpret our experiences through our conceptual systems; thus, we exclusively get in contact with the aspects of reality through this framework. (Polanyi 1958, 196-197) At the same time, the states of contemplation, submersion, and dwelling in lead us to the sacrifice of our own individuality and, as a result, to getting into a closer relationship with things.

> *"Correspondingly, the impersonality of intense contemplation consists in a complete participation of the person in that which he contemplates and not in his complete detachment from it, as would be the case in an ideally objective observation."* (Polanyi 1958, 197)

In this regard, the objectivist Laplacian ideal of knowledge has valid foundations; however, objectivist thinking presents it inaccurately: a transcending subjective point of view does not have its roots in distancing ourselves from, but instead in forming a closer relationship with things and in contemplation—namely, in the condition that Polanyi calls dwelling in. Denying the relevance of direct connection with things and accepting the objective exclusively, distancing the way of cognition as relevant, leads to alienating ourselves from reality and to the intrinsically meaningless world of objective details, just as Polanyi explained it several times with regard to the Laplacian ideal of science. (Polanyi 1958, 199)

Conclusion

In this chapter, Polanyi's most important claim is that even the highest levels of articulated systems have tacit roots that go back to the beginnings of animal life. Consequently, in exclusively explicit, impersonated form, these systems do not mean anything—this would be the knowledge according to the Laplacian ideal—because they obtain their meaning owing to those anciently rooted tacit foundations upon which they lean in the course of their operation.

The tacit capabilities we share with animals—e.g., perception—develop into intellectual passions and exceptional abilities in the higher levels of human conceptual and cognitive schemes. Such ability is, for instance, the recognition of intellectual beauty, which is inaccessible for animals and even for insufficiently sophisticated human beings in spite of their collective tacit roots. Tacit foundations and articulated conceptual systems are in dynamic correlation with each other.

Polanyi distinguishes private and public passions from each other. Private passions are of an animalistic nature and pursue the completion of the actual desires of a given person. In contrast, intellectual passions perpetuate themselves by their own completion, provided by the higher level of the articulated conceptual and explanatory system in the framework they operate and which enables them to develop. This same scheme calls to life the culture-bound institutions—e.g., science—that support and cultivate the development of intellectual passions. In contrast with animal passions, intellectual passions serve not only the development—and in correlation with that, the completion of—the (intellectual) needs of a given person but do the same with a given community.

According to Polanyi, even the simplest perception is connected to a theory of universe that enables us to take hold and evaluate the aspects of reality. A cat's tacit conception about a frog is obviously far from that of zoologists. The tacit active principle, however, which ultimately directs the comprehension of both the cat and the zoologist, originates from the same shared tacit root. In these thoughts, we can note Polanyi's committed evolutional point of view and emergent ontology (see ch. 13).

Chapter 7

Conviviality

Goals of the chapter

1. According to Polanyi's main declaration, the best form of society is a free society since it acknowledges and supports intellectual passions, which are indispensable for moral, cultural, and scientific development.
2. He argues that only such a society is able to set the community up with sophisticated articulated systems such as science that feed and satisfy intellectual passions. For that very reason, a free society is obligated to support the satisfaction of intellectual passions; namely, it has to provide its community of scientists with the financial, social, and institutional conditions necessary for free thinking and research.
3. He introduces trust as a key element of a free society. According to Polanyi's description, trust has several roles. On the one hand, it is necessary in order to sustain knowledge since it enables persons to consider a universal statement to be true while knowing that it can, in fact, be wrong. On the other hand, trust is fundamental in submission to authority, which Polanyi considers indispensable for learning and for the operation of diverse social institutions.
4. He introduces the four elements of his vision of society: culture, as a portion of collective convictions; community, as participation in collective activities; economic cooperation; and executive power, as the application of authority and compulsion, all of which he considers indispensable for the operation of a righteous system.
5. He describes the concept of *moral inversion*, which marks the psychological condition of a human denying the existence of intellectual and moral passions while these passions still operate within, causing them to reach their fulfillment in the denial of their own existence. Since moral inversion can never manifest in its fullness, it is, rather, a psychological and social process that gradually strengthens. Polanyi also discusses the disguised forms of moral inversion.
6. He considers faithfulness to the truth to be the most important obligation for a scientist, and he views it as the central value of science, which in a social context means that a scientist is faithful to free society since it is the one that respects truth and various intellectual values.

***

The concept of conviviality and the tacit foundations of transmission of social lore

The development of knowledge in different sciences is granted by the tacit foundations, which sustain the operation of articulated systems of knowledge.

> *"This tacit sharing of knowing underlies every single act of articulate communication. I shall then take in the whole network of tacit interactions on which the sharing of cultural life depends, and so lead on to a point at which our adherence to the truth-can be seen to imply our adherence to a society which respects the truth, and which we trust to respect it."* (Polanyi 1958, 203)

Articulated communication is made possible by commonly possessed tacit knowledge, which motivates and fills the acts of explicit communication with meaning (in detail see ch. 8) and from which these acts of explicit communication emerge. This shared tacit knowledge, which we rely on when evaluating explicit manifestations, is the same in everyone. It is the only way—due to the lack of any other foundation—to create communication based on mutual understanding. Tacit knowledge, however, is not exclusive to man.

In the case of animals, tacit knowledge, of course, cannot be considered as the basis of articulated communication as they are not even able to articulate, but the transmission of knowledge has its completely tacit, prearticulation forms from which articulated communication has emerged. Polanyi mentions that although bees also communicate, their communication does not result in individual discursive thinking. Bees do not apply different symbols as instruments of motivating intellectual development but exclusively as general instruments of their present communication. (Polanyi 1958, 205) According to Polanyi's standpoint, tacit knowledge "is transmitted on a primordial level from one generation of animals to the next by an imitative process which students of animal behavior call *mimesis*." (Polanyi 1958, 206) However, in the case of animals, mimesis is rather limited and closely linked to instinctive activity. (Polanyi 1958, 206)

Polanyi calls the coexistence and communal life of people with shared tacit knowledge *conviviality*. Upon introducing the concept, he mentions two examples. The first is the case of children playing together who develop such close companionship during their play that they insufficiently realize the distinction between themselves and their playmates. (Polanyi 1958, 206-207) His other example is the similarly close relationship of animals living in a flock or herd.

> *"A true transmission of knowledge stemming from conviviality takes place when an animal shares in the intelligent effort which another animal is making in its presence."* (Polanyi 1958, 206)

According to Polanyi, the transmission of knowledge on a prearticulation level is clearly perceptible in chimpanzees, but we can observe the same in less developed mammals and birds too. Mimetic reactions of offspring towards the older animals feeding and protecting them are similar to the tacit language-learning actions that children present while acquiring articulated speech as a craft. Accordingly, the roots of articulated speech are found in the tacit communications of animals. This is indicated by the tacit, craft-like nature of articulated communication and the fact that articulated sentences obtain their meaning upon their tacit foundations (see ch. 8). By articulation, the tacit capabilities have been equipped with intellectual instruments that, in the case of humans, enable their operation to exceed similar tacit capabilities of animals to a great extent (ch. 5).

> *"Spoken communication is the successful application by two persons of the linguistic knowledge and skill acquired by such apprenticeship, one person wishing to transmit, the other to receive, information."* (Polanyi 1958, 206)

Thus, a precondition of communication is the kind of conviviality in which authority and trust tacitly connect the participants in communication.

In terms of its fundaments, the transmission of social lore occurs in exactly the same way as the learning of a language—that is, social knowledge is not impersonal and not exclusively explicit since it is operated by the tacit foundations, which, due to their craft-like characteristics, are only obtainable by standing for conviviality and accepting the institution of authority and advancing trust. (Polanyi 1958, 208) Transmission of social knowledge is, therefore, built on the authority and trust between participants, who in this case are master and apprentice. This kind of transmission also happens within the framework of much more sophisticated social systems and institutions than everyday communication between people. (Polanyi 1958, 207) This process is only operable if the apprentice joins the community in advance, before obtaining knowledge, and acts according to its standards. Joining a community and standing for conviviality means offering one's personal faithfulness, which is an act of subordination, by accepting the authorities and values of the given community.

Polanyi notes that, due to their nature, these self-transforming processes are informal, irreversible, and acritical. One of his arguments is that a significant part of knowledge incorporated in sciences is of a communicative nature; i.e., it is not composed of the results of our own observations. Accepting any knowledge of a communicative nature, however, is only possible through tacit actions based on authority and trust. (Polanyi 1958, 208-209)

Receiving a new member, e.g., an apprentice, into the community, however, always causes a tiny, often unnoticeable transformation in the prevalent consensus since the newly arrived person always adds his own ideas and interpretation into the accepted consensus. Actually, he does not even accept the prevalent consensus, but the one he considers prevalent, according to his own tacit convictions. This process is similar to when the meaning of a word changes slightly upon using it in a new situation. (Polanyi 1958, 208)

If there is no collective tacit foundation, there will be no collective communication either. Those who attempt to change a consensus are always parts of this consensus, although they have different ideas about some of its explicit details. Despite any semblance, in the course of such scientific arguments, "dissent does not seek to abolish public authority, but to claim it for itself. (Polanyi 1958, 209) Namely, it wants to bestow certain details with new meaning and thus tries to change the prevalent consensus to a certain extent.

Conviviality draws most of its effectiveness from common experience and conversation. According to Polanyi, transpersonal communication between people is driven by the desire for company instead of rational considerations or interests. Accordingly, clean conviviality and maintenance of good relationships direct several acts of articulated communication, making the cooperation of the group members more effective. (Polanyi 1958, 210)

Diverse ceremonies and various collective forms of festive attendance also serve the maintenance of clean conviviality.

> *"By fully participating in a ritual, the members of a group affirm the community of their existence, and at the same time identify the life of their group with that of antecedent groups, from whom the ritual has descended to them."* (Polanyi 1958, 211)

Polanyi notes that various festivities and rituals are "a celebration of convivial existence," and thus they bring about hostility from several modern critical perceptions, which can range from fundamentally denying any value for group life, to being unable to associate similar events in corporate life with any palpable benefit, or, even worse, being convinced that rituals and festive gatherings suppress individuals' genuine, spontaneous sentiments. (Polanyi 1958, 211)

Our intellectual passions are neither directed towards themselves nor others since their operation is assigned by *universal intentions*. Consequently, they create comprehensive, collective systems of values that are closely linked to other fundamental systems of values within human communities, such as law, habit systems, or morality. The collective-value horizon and the acknowledgment of the institution of authority mean the undertaking of certain collective obligations for the participants of convivial existence. (Polanyi 1958, 212) As a result, the group expects a conformal attitude from its members.

> *"This acknowledges a common good for the sake of which deviation may be suppressed and individuals be required to make sacrifices for defending the group against subversion and destruction from outside."* (Polanyi 1958, 212)

According to Polanyi, moral judgments are closely connected to intellectual passions because, in contrast to animal ones, moral and intellectual passions are *public passions*. (Polanyi 1958, 215) Moral judgments, however, are taken more seriously than intellectual evaluations. Man's moral power and deeds are not evaluated as outward performances as is the case, for example, with scientific discovery, but rather by the effect they have on the whole of our personality.

> *"Accordingly, moral rules control our whole selves rather than the exercise of our faculties, and to comply with a code of morality, custom and law, is to live by it in a far more comprehensive sense than is involved in observing certain scientific and artistic standards. […] their lives are ordered by the same morality, custom and law, which jointly constitute the mores of their society."* (Polanyi 1958, 215)

Polanyi distinguishes the individual practice of social knowledge when society encourages individual thinking and another kind of practice of social knowledge in which we reconcile our own life with all this. Society calls the latter "an administration of society in accordance with essentially civic thought." (Polanyi 1958, 215)

> *"But this dependence will enter differently into the two types of thought in society, since civic culture itself sustains the civic institutions of society, while individual culture is, on the contrary, itself sustained by these institutions."* (Polanyi 1958, 215)

In terms of its essence, however, civic culture is sustained by intellectual passions.

Polanyi distinguishes four areas that compose any system of fixed social relationships. (Polanyi 1958, 215)

1. *Culture*: A conventional way of articulated completion of intellectual passions is the *sharing of convictions*. The main scenes of culture are universities, theaters, etc.
2. *Communities*: A "dwelling in" way of articulated completion of intellectual passions is the *sharing of a fellowship*. Its main scenes are associations, parties, and the army. These are the institutions of conviviality that aim to strengthen group loyalty.
3. *Economic system*: Its most important characteristic is *cooperation* for a common material purpose. Its main scenes are the diverse economical corporations.
4. *Public power*: Serves the protection and direction of the other three institutional systems. Its operation is supported by *authority and coercion*.

Static and dynamic societies

Polanyi's perception is that, despite all greater reforms, the hierarchic social structure had been considered fundamental until the American and French Revolutions and the idea that society—regardless of its present conditions—is always able to develop became accepted only after these. Until this point, we talk about *static*, from here about *dynamic* societies. The difference is how the future is perceived and visualized in certain societies and what is considered an appropriate social act regarding the future of that society.

There are two types of dynamic societies: *revolutionary*, which is totalitarian since, according to the Laplacian program, all thoughts are subordinated to welfare; and *reformist* or free society, "which accepts in principle the obligation to cultivate thought according to its inherent standards." (Polanyi 1958, 213) In static societies, thought is also recognized as individual force, and truth is respected just as in free society, but since intellectual life operates within strict limitations, we cannot talk about freedom of thinking within static societies. (Polanyi 1958, 213)

Table 7.1: *Fundamental characteristics of static and dynamic societies*

Static societies	**Dynamic societies**	
Purpose: conservation of social conditions	Purpose: perfection of society	
Strict social limitations	No strict social limitations	
	Revolutionary	**Reformist**
Truth is respected	Truth is not respected	Truth is respected
No free thinking	No free thinking	Free thinking

Intentions of social perfection in dynamic societies ensue from the free thoughts and ingenuous feelings that destroyed the static, authoritarian social systems of the medieval ages. This historical process, however, created a hazardous contradiction in the emerging free societies, which sometimes led to the development of totalitarian regimes. (Polanyi 1958, 214) The reason for this is as follows:

> *"The great movement for independent thought instilled in the modern mind a desperate refusal of all knowledge that is not absolutely impersonal, and this implied in its turn a mechanical conception of man which was bound to deny man's capacity for independent thought. Such objectivism must represent the public good in terms of welfare and power and set in motion thereby the self-destruction of freedom. For when open professions of the great moral passions animating a free society are discredited as specious or Utopian, its dynamism will tend to be transformed into the hidden driving force of a political machine, which is then proclaimed as inherently right and granted absolute dominion over thought."* (Polanyi 1958, 214)

Modern critical thinking thus not only liberated later modern societies from the strict medieval conceptual and institutional limitations but at the same time challenged the existence of tacit foundations, intellectual, and moral public passions that are the fundaments of a free society and free thinking. As a result of this, referring to the moral dynamism of modern societies, social development, and welfare, certain political forces tried to eliminate the concept of freedom, free society, and free thinking. An apparent example of this process is Marxism and the Soviet systems.

In the case of the static-dynamic pair of concepts, Polanyi's approach is strongly simplifying since he does not mention the consecutive steps of the transformation process at all. Thus, we should consider the static-dynamic pair of concepts as the two endpoints of the transformation process. While in the fifteenth century, we can only see static societies, in the nineteenth century, we can unambiguously speak about dynamic societies. The final, decisive moments of the transformation process were the American and French Revolutions. However, the transformation had already started in the sixteenth century, and we can consider the societies that came to life as a result of revolutions in England and the Low Countries (Netherlands) as having been partly dynamic already in the seventeenth century.

Moral dynamism of science and free societies

According to Polanyi—in contrast with the generally prevalent concept—we do not acknowledge the scientific consensus because its affirmations are built on observations that can be repeated again by anyone, anytime. It is exclusively an objectivist declaration of our trust in science and our insistence for consensus. Most of the observations are never repeated. If there was an opportunity to reaffirm and we actually repeated the observations, just by themselves, they could not confirm the commitments in whose development the observed facts played an active role. The reason for this is that it cannot be exclusively confirmed as to why these facts were chosen rather than any other facts (see ch. 6). (Polanyi 1958, 217-218)

Besides the scientific consensus, there is also a social consensus surrounding science, which manifests when society becomes receptive to the intellectual passions of the scientific elite. As a result of this, society both financially and morally supports scientific institutions and the propagation of scientific theories and results. (Polanyi 1958, 219)

Social sciences, arts, and religion have a similar role as natural sciences in a free society. There is one important difference, however. In these domains, there is much greater diversion among ideas than in natural sciences. Today, "new" holds an indisputable value, owing to the cult of innovation, as does artistic genuineness—according to Polanyi's point of view—due to its personal nature. Novelty can cause a much more comprehensive perceptional change than what is usual in social sciences, but significantly less perceptional change occurs in natural sciences in general. (Polanyi 1958, 220) In our educational system, only natural sciences are situated on a firm foundation. Any other cultural authority has to fray out its own recognition, often even with the other scholarly institutions of its own sphere of profession. (Polanyi 1958, 221) In a free, pluralistic society, the rival leaders directing the operation of diverse schools of science share a similar status. (Polanyi 1958, 221-222)

> *"This belief in an autonomous process of coherent thought is (as in science) the fundamental condition for the social cultivation of thought, guided by its own standards and prompted by its own passions."* (Polanyi 1958, 222)

This means that we recognize and accept intellectual passions in diverse articulated forms and values because we believe in the independent value of thought. We do not want to subordinate its operation to any non-scientific consideration, as in the case in static societies, but nor do we want to subjugate it to utilitarianism or want it to become a mere instrument of financial welfare.

Although Polanyi heavily criticizes critical philosophy in several parts of his book and draws attention to its serious dangers that threaten free society (see next section), he affirms beyond dispute that the nineteenth and twentieth centuries led to the unique development of civic culture, directing society towards moral perfection. Polanyi mentions the humanization of penal law and execution, army, naval fleet, schools, and mental institutions. Significant changes took place in the family and in the social sphere; industrial laws created better working conditions; institutions of care were established for the old, the sick, and the disabled; elective rights were extended to an even greater population of society; trade unions were recognized; discriminative rights for women, Jews, and Catholics were abolished, etc.

> *"Such is the dynamism of the modern free society. It consists in the moral progress of civic thought, which transmits its conclusions, through the machinery of self-government, into acts of social reform. It is the practical outcome of an intellectual process, moved by its own passions and guided by its own standards."* (Polanyi 1958, 223)

The point of the change is tacit; it appears later in explicit laws and in the operation of social institutions. Thus, in the beginning, we can talk about the gradual change in ways of social contact and family customs, and in connection with all that, in informal moral judgment. (Polanyi 1958, 223)

> *"We may take it, then, that in an ideal free society civic life would be continuously improved solely by the cultivation of moral principles."* (Polanyi 1958, 223-224)

Moral inversion

Moral inversion is a peculiar consequence of critical philosophy and the moral dynamism of free societies appearing in dynamic revolutionary societies.

Polanyi considers the phenomenon of moral inversion most expressively describable and analyzable in the case of communist movements. However, he mentions Friedrich Meinecke's interpretation that German power politics led to the First World War. According to Meinecke, the Germans first understood the original sinfulness of power, and upon making power and law equal, they declared that any other behavior was just a sign of false morality. In other words, by denying morality, they tried to morally justify their deeds, pleading raw sincerity—with moderate success—while criticizing the false morality of Anglo-Saxons, which they tried to disassemble; thus, in fact, trying to gain a moral advantage. This same contradiction appears in Marxist thinking and acts: they try to gain a moral advantage by denying morality while providing a much more detailed ideological background for actions. (Polanyi 1958, 226-227)

Moral inversion is based on a determinative contradiction of two factors that we can perceive in modern ages. Namely, in a free society, there is a strong tension between the pursuits for moral perfection within society (moral dynamism) and the Laplacian ideal of knowledge, which, due to critical philosophy, has become deeply rooted in modern science. This concept of knowledge requires the perception of humanity and human society through the lens of strict objectivity, thus denying the existence of morality. Polanyi calls the former factor *dynamic* and the latter *objective*. The convincing power of Marxism rises from its ability to resolve this tense contradiction.

> *"[Marxism] enables the modern mind, tortured by moral self-doubt, to indulge its moral passions in terms which also satisfy its passion for ruthless objectivity."* (Polanyi 1958, 228)

Polanyi does not have the slightest doubt that man possesses real moral passions; thus, objectivism is a false doctrine. Namely, man is not a mechanical machine but a sentient, passionate living being. Those who deny the existence of moral passions and morality in the spirit of scientific objectivism also do so upon being driven by moral passions.

This contradiction can lead to serious internal instability and confusion if one tries to address the problem. If we were able to find a doctrine, the conceptual—and thus cognitive—structure of which indicates that moral passions disguise themselves as seemingly unrelated to morality, this contradiction, which previously tormented the individual, would suddenly cease to exist. Disguised moral passions, however, would express their effects in an unseen way. This is exactly what happens to the followers of Marxist doctrines. This opens wide a gate towards the moral dynamism of modern man, but at the same time does not have to address the problem of philosophical and scientific verification of moral judgments.

Marxism obtains this complicated unity among moral dynamism and scientific objectivism with a rather simple intellectual action. Namely, it states that the moving force behind historical events is necessity; thus, the utopian society of communism, or "paradise on Earth," is going to materialize thanks to the inevitable development of production fuelled by technology and modern science. Polanyi highlights that this doctrine follows the same logic as any other primitive method that presumes that there is a particular authority or force behind natural and social events, which can be either magical or not, divine or profane. (Polanyi 1958, 229) The concept of historic necessity is a guarantee of moral development in modern society. The utopia of communism is actually the morally perfect society described with strictly objectivist ideas and foretold by mechanistic actions in Marxism. (Polanyi 1958, 229)

Furthermore, Marxism makes an offensive against civic ("bourgeois") moral endeavors, unmasking them as manifestations of exclusively financial and hegemonic interests. The formulation of this accusation, however, bestows Marxist doctrine with further (disguised) convincing moral power while denying its independent existence and verifiability, and thus the possibility of moral judgments. (Polanyi 1958, 229) According to the self-image of Marxism, it would be similarly hypocritical to set out such "naïve" goals as the implementation of freedom, fraternity, and righteousness. However, since it defines itself as a scientific idea, it formulates its goals within the framework of scientific discourse. Equality and welfare can only come to fruition upon the dispossession of the means of production by the proletariat, i.e., a victory in the class struggle. This doctrine, however, cannot be grounded with instruments of science. But its convincing power is not even built upon that; the source of its convincing power is its use of (morally neutral) scientific concepts to define goals that are actually of moral nature.

> *"By covering them with a scientific disguise it protects moral sentiments against being deprecated as mere emotionalism and gives them at the same time a sense of scientific certainty; while on the other hand it impregnates material ends with the fervor of moral passions."* (Polanyi 1958, 230)

We can see that, while Marxism unmasks and denies moralism, tacitly, it still builds on the power of moral endeavors. (Polanyi 1958, 230) Polanyi calls this structure *dynamic-objective pairing.* The first term refers to the *moral dynamism* of modern society, and the latter to the scientific *objectivity* of the Laplacian ideal of knowledge.

"Any criticism of its scientific part is rebutted by the moral passions behind it, while any moral objections to it are coldly brushed aside by invoking the inexorable verdict of its scientific findings." (Polanyi 1958, 230)

The convincing power of Marxism thus originates from its defining itself as a scientific theory, and in this context, using the objectivist language of science and the mechanical concept of man. At the same time, however, it sticks the utopian goals of radical perfection of society on its flag, channeling the moral dynamism of free modern society into Marxist political movements. Several ideological approaches, as well as numerous political movements, criticize the present deficiencies of civic society; consequently, the genuineness and convincing power of Marxism does not originate from its reform-purposed criticism but rather from the above structure of moral inversion.

Moral inversion is dangerous because moral passions and judgments go through fatal transformation if they are represented as scientific statements, thus becoming isolated from their fundamental moral context. In such cases, moral passions tend to become animalistic due to the lack of a conceptual system articulating morality and the institutional system supporting the expressions of moral passions; as a result, their articulation becomes limited by the framework of the objectivist conceptual system that even denies their actual existence.

Moral passions start to become animalistic because they are built on tacit, animal roots of convivial existence that can be experienced as ingenuous feelings only within the conceptual systems of articulated human morality in the framework of the institutions exhorting their development. However, the process pointing towards these ingenuous sentiments is cut, and as a result of this, moral passions run back wildly to their animalistic roots. This leads to fanaticism and a cult of power since, in this case, the common good can only be expressed in material concepts in correspondence with the mechanical image of man in science, so these distortions lead to the development of modern totalitarian society. The ingenuous sentiments and endeavors attached to the passions invigorating the moral dynamism of modern society cease; political power, in turn, becomes self-serving. (Polanyi 1958, 231)

Polanyi runs a parallel between Stalin and Hitler, highlighting that Hitler profited a lot from the Marxist-Bolshevik example, but his national-socialist movement was rooted primarily in German romantic nihilism.

"This doctrine taught that an outstanding individual is a law unto himself and may, as a statesman, unscrupulously impose his will on the rest of the world, and that a nation has likewise the right and the

> *duty to fulfill its 'historic destiny' irrespective of moral obligations."* (Polanyi 1958, 232)

Polanyi later describes the phenomenon of modern cultural nihilism, which, according to him, arose as a result of the joint effect of the romantic and modern scientific movements. (Polanyi 1958, 235) Moral nihilism is the mark of exceptionally strong moral passions, and its followers tend to sympathize with Marxist and national-socialist revolutionary movements. (Polanyi 1958, 236)

A logically sounder, more consistent standpoint of moral inversion can come to existence through deceptive substitution. Polanyi speaks about the *disguised forms* of moral inversion in such cases in which the inversion has not yet become coherent moral inversion yet. In the latter case, inversion becomes part of *political practice*, while in the case of disguised forms, it appears only in theory within *public or scientific thinking*, but everyday life is still organized along with traditional moral ideas and endeavors. (Polanyi 1958, 233) According to this:

> *"Men may go on talking the language of positivism, pragmatism, and naturalism for many years, yet continue to respect the principles of truth and morality which their vocabulary anxiously ignores."* (Polanyi 1958, 233)

Among others, Polanyi mentions the example of Freud, who traces back the moral life of man to psychological factors other than morality; however, he is still committed to an ingenuous society that seeks the prosperity of all; in other words, he disguises his moral motivations in scientific language. As we can see, Freud did not want to enforce his scientific concepts in political practice. (Polanyi 1958, 234)

> *"Indeed, a writer who has proved his hard-headed perspicacity by denying the existence of morality will always be listened to with especial respect when he does moralize in spite of this."* (Polanyi 1958, 234)

Due to its structure, this scientific conceptuality disguising morality can even operate successfully against moral nihilism. Polanyi mentions the works of two outstanding figures of pragmatism and utilitarianism, Dewey and Bentham, as examples. (Polanyi 1958, 234)

This, however, does not mean that Freud or other great thinkers would deliberately misguide their audiences; it is simply how moral—in this case disguised—inversion works. Ensuing from the Laplacian ideal of knowledge, we deny the reality of moral judgments; however, our pronouncements of objectivist language tacitly impregnate with serious moral content—which is a typical case of deceptive substitutions (see ch. 1), and, at the same time, the antechamber of moral inversion.

From the eighteenth century, mostly because of utilitarianism, disguised forms of moral inversions became more and more determinative; however, they only managed to pervade public thinking in the twentieth century, thus leaving moral judgments theoretically defenseless. The traditional articulated conceptual schemes and the institutional systems—which enabled the expression of moral passions and corrected them when necessary—have been discredited. Only such utilitarian and materialist conceptual systems—and the scientific institutions accepting them—could remain accepted in a society which were not capable of cultivating moral passions and thus of properly supporting our moral judgments. Specifically, this means the discrediting of Christian denominations, Christian faith, and, in connection with that, the traditional ethics laid upon Christian values. These are the institutionalized forms of Christianity that, until the twentieth century, were able to successfully fulfill the social functions traditionally bestowed on them in free modern societies.

According to Polanyi, the recognition of moral inversion has to mean the admission that our fundamental motives are of a moral nature, as it also ensues from our Christian image of humanity. This admission, however, does not mean at all that we should return to the medieval static Christian social structure. Polanyi is fully committed to the modern, dynamic, and free society. According to his suggestion, there is a need for the application of a new, so-called post-critical philosophical and scientific language, and, at the same time, for the rooting of a frame of mind, that is able to subtract the consequences ensuing from this recognition and thus develop a new scientific worldview with a befitting social system of institutions.

Polanyi highlights that, although modern skepticism against any kind of belief grew as strong as it did in the twentieth century, it could not affect science. Belief in science remained unquestionable even for the Marxist since critical philosophy does not recognize the tacit foundations of knowledge, and thus the fiduciary foundations of science either; as a result, for critical philosophy, the abolishment of any belief just amplifies the belief in science. (Polanyi 1958, 238) This critical conviction only started to change after the completion of *Personal Knowledge*, with the spread of relativism and the detailed research conducted in the history of science in the sixties and seventies.

On the grounds of the Marxist interpretation of science, the unmasking of so-called bourgeois science and the abolishment of the independence of science had already begun before Polanyi stated that every step in the progress of science occurred in response to practical needs. This unmasking process was directed towards those foundations which provided the individual and institutional conditions to the free operation of science, identifying them as selfish interests. In the Marxist interpretation of science, social and economic factors become exclusive, thus denying the free practice of science for its own sake. In the name of the proletariat, science became restricted to serve social welfare—or by another, more truthful interpretation: the Party. (Polanyi 1958, 238) However, this attitude to research, which ensued from the Marxist image of science, was not practically applicable, and a decent number of "red tails"—which is a practical case of backward unmasking—provided the mostly free practice of science.

The changing position of liberalism

In the nineteenth century, liberalism, as a leading political doctrine of a modern, free, dynamic society, fought against the remains of the old static system and for the reasonable moral amelioration of society.

> *"But when the appetitive conception of man denied the reality of moral motives in public life, the ideals of liberalism were inverted into the doctrines of modern totalitarianism."* (Polanyi 1958, 244)

In the nineteenth century, liberalism was not only able to mask its moral endeavors with diverse deceptive substitutions based on an objective, mostly utilitarian set of concepts—disguised forms of moral inversions—but even gained some more persuasive power from these versus the so-called power of the past. In the twentieth century, however, moral inversion has developed into a logically more sound condition through the use of the set of concepts and doctrines of liberalism. One example is the Marxist communist movement, which turned over the originally ingenuous, although masked, liberal moral endeavors into immorality, thus weakening the moral power of liberalism as a doctrine of political philosophy and a pervading idea of social movements. By Polanyi's times, the position of liberalism became highly unstable since it had to face the fact of finally becoming unable to build its moral endeavors on the scientifically exact foundation borrowed from critical philosophy. It could only lean on the traditional institutions and the private moral convictions of people; therefore, according to Polanyi, it had to become conservative to avoid moral inversion and totalitarianism. (Polanyi 1958, 244)

> *"The recognition granted in a free society to the independent growth of science, art and morality, involves a dedication of society to the fostering of a specific tradition of thought, transmitted and cultivated by a*

> *particular group of authoritative specialists, perpetuating themselves by co-option."* (Polanyi 1958, 244)
>
> *"Unjust privileges prevailing in a free society can be reduced only by carefully graded stages; those who would demolish them overnight would erect greater injustices in their place."* (Polanyi 1958, 245)

Although led by his own philosophical considerations, Polanyi essentially takes a standpoint similar to that taken by Edmund Burke in his definitive work entitled *Reflections on the Revolution in France*: basic liberal values, moral dynamism of modern free societies, and freedom of science can only be saved from the totalitarianism-motivated new political movements by a *conservative turn.*

Conclusion

According to Polanyi, human comprehension and transmission of knowledge are based on those tacit foundations that are identical in terms of their roots to the tacit capabilities of animals. The articulated conceptual and explanatory systems and public institutions, being responsible for the deployment and guidance of moral and intellectual passions, are deeply embedded in these tacit passions. Therefore, tacit moral and intellectual public passions, although only indirectly but still fundamentally shape human societies.

Societies that exclusively pursue the preservation of existing social institutions are called static societies. These respect truth and consider intellectual and moral values important but do not pursue the development of society. In these societies, thinking is forced between strict conceptual and social limitations; consequently, we cannot consider them free. In contrast to the static society, one which pursues the perfection of its institutions is called dynamic. Since the moral life of the community lies on tacit foundations, its development manifests mostly in the deployment and progression of intellectual and moral passions.

A dynamic free society, alongside moral development, also aims for the deployment and completion of intellectual passions that are related to moral passions. This means respecting the truth and recognizing thought as an independent power that can set up the theoretical foundations of the development of social institutions by its achievements.

However, in modern societies, an objectivist-mechanical perception of science became dominant that at first challenged and then started to deny the existence of intellectual and moral passions. This resulted in the practice of hiding the discourse over the moral perfection of society in the morally neutral conceptual set of objectivist-mechanical frameworks that create the disguised forms of moral inversion. Consequently, significant tension has formed among the theoretical reflection concerning science and the social practice of employing its results.

A logically sounder outcome of this is *moral inversion*, which led to the revolutionary emergence of modern totalitarian societies that try to resolve the tension mentioned above by converting the philosophical-scientific theory into practice. It described—or, rather, foretold—the utopian communist idea of the moral perfection of society and the ways of its implementation in a morally neutral language to conform with the prevalent scientific approach. It secured its convincing power borrowed from scientism by channeling the moral dynamism of modern societies into the movements emerging on behalf of the revolutionary theory.

However, neither conceptual and explanatory systems like this nor the social institutions built upon them are capable of maintaining and deploying intellectual and moral life, namely, intellectual and moral public passions. Revolutionary totalitarianism ultimately leads to the emergence of an immoral power structure and the abolishment of free thinking. Science in its entirety is forced to serve material welfare and the totalitarian power itself.

Part Three: The justification of personal knowledge

Chapter 8

The logic of affirmation

Goals of the chapter

1. Polanyi shows that any affirmation necessary includes a tacit component: a purely objective scheme is meaningless (subch. 1-7).
2. He argues that systems based on logical deduction cannot model the workings of the human mind because such systems exclude the tacit component. These systems are merely tools of real human inferences; they are not the same with them (subch. 8-11).
3. He puts his conclusions and then the whole book into a historical context (subch. 12). He calls his endeavor *fiduciary program.*

It is important to emphasize that, although the main goal of the chapter is to establish a new philosophy of language based on the concept of tacit knowledge, the last subchapter goes far beyond this goal and tries to establish the aim of the whole book, and thus the real aim of philosophy according to Polanyi.

***

The tacit nature of affirmation

Polanyi claims that every articulate assertion has two composite parts: 1. *sentence*; 2. *tacit act.* The sentence is an *explicit formula.* The tacit act is a *commitment* by which a person asserts the sentence. It follows that the explicit sentence gains meaning and refers to an aspect of reality *only* by the person's tacit commitment. (Polanyi 1958, 254) This can be depicted in the following way:

Explicit sentence + Tacit commitment → An aspect of reality

> "*[...] only a speaker or listener can mean something* by *a word, and a word* in itself *can mean nothing.*" (Polanyi 1958, 252)

The sentence "snow is white" means nothing in itself. This means anything and becomes true—or in the case of "snow is red," it becomes false—if and only if a person refers with it to an aspect of reality. Without this reference, both sentences are only vacant explicit phrases. It follows that referring to reality is not another explicit sentence; otherwise, nothing would change with the original "snow is white" or "snow is red" sentences, but it is a *tacit commitment* that, according to our beliefs, reality is in a kind of state; that is, snow is white or red. This act of commitment is not explicated; it is simply not part of the explicit sentence "snow is white."

Therefore, if a person is not committed to the content of a sentence when he says it, it remains a vacant explicit sentence and does not become an assertion about reality. The phrase "snow is red" is not an assertion, just a sentence unless someone really believes that snow is red. Consequently, a lie is also not an assertion. Of course, someone can be wrong, but then he has to assert that snow is red, honestly. (Polanyi 1958, 254) So a sentence can be part of an assertion or not, and, based merely on the explicit form, it cannot be decided which is the case. It can be decided *based only on the tacit context.* Thus, in Polanyi's philosophy of language, semantic examinations cannot be conducted by merely investigating explicit sentences.

Sentences of an encyclopedia or a scientific paper in themselves have no meaning at all either—no matter how precisely and exactly they were explicated. They gain their meaning if and only if somebody asserts them according to their tacit convictions and previous experiences. It follows that two persons do not and cannot understand exactly the same from an explicit sentence. This could only happen if they have exactly the same tacit convictions and previous experiences.

Polanyi differentiates between the *confident,* or direct, and the *skeptical,* or oblique, use of a word or a sentence. According to the former case, someone asserts a sentence honestly with commitment; to the latter one, he suspends his assertion to examine its (explicit and tacit) contents. (Polanyi 1958, 250)

Polanyi based his conclusions about the nature of assertions on the results of previous chapters (mainly on ch. 4 and 5). According to these results, articulation and language are not self-standing or based on some mental principle but based on skills; that is, a person can articulate his thoughts, beliefs, feelings, etc., merely by his tacit bodily skills. After acknowledging the evolutionary origin of human beings, it can be considered as trivial, but the consequences are not trivial at all: *tacit knowledge is the precondition of all articulate, explicit sentence and thought*; that is, every assertion is based on tacit acts.

This is Polanyi's starting point in this chapter, and the problem he is grappling with is the following: typically, modern sciences try to describe all aspects of reality as explicitly as possible; e.g., in the case of a person's act, using the more explicit neurological and physiological descriptions. This is true, of course, for all cognitive processes in the mind too. However, if Polanyi's examination of the logic and tacit nature of affirmation is well-established, it is a total misunderstanding of these processes.

Since any affirmation is not just an explicit sentence but also a tacit act, it cannot be asserted, only referred to in the denotative sense by another descriptive sentence, or it can be done again—that is, it can be confirmed with another tacit act in the existential sense (see ch. 4). This means that an act can*not* be asserted, only an explicit sentence that describes an act. "Snow is white" in itself is a sentence. "Snow is white" is, however, an affirmation, a tacit act referring to reality because the talking person is committed to its truth. This is the reason that it has a denotative meaning.

It follows that "snow is white is true," contrary to "snow is white," is not just a vacant explicit sentence in itself but a declaration of a person's commitment. Therefore, if somebody completes an explicit sentence; e.g., "snow is white" with the phrase "is true," then the original sentence is no longer just a sentence that can be asserted or not as somebody wishes, but necessarily an assertion, which was at once confirmed by the word "true."

The main consequence of this is that it is entirely meaningless to ask whether "snow is white is true" is true or not, because it is not a sentence that can be asserted but a personal commitment to an aspect of reality, just like the *tacit act* of riding a bicycle. Riding a bicycle is a different type of personal act from confirming an act of affirmation with the word "true," but both are personal acts, which is why they can be neither true nor false. It is entirely meaningless to ask whether Peter's bicycle riding is true or false because it is an act that, according to the nature of acts, is *successful* or *unsuccessful* but not true or false. One can only ask whether the sentence "Peter is riding a bicycle" is true or not—if, of course, somebody asserts it; and this assertion will be true or false depending on whether Peter rides a bicycle successfully or not. (Polanyi 1958, 255)

So Polanyi claims that there are *two significantly different types* of linguistic expressions. The first type includes those normal expressions by which a person can compose sentences and assert something. The second includes those linguistic expressions that confirm or reject the use of expressions of the first type; that is, they do not compose sentences but realize acts—with a well-known expression, they are speech acts. (Polanyi 1958, 255)

BOX 17: Speech acts and the logic of affirmation

There is an obvious similarity between Polanyi's philosophy of language and J. L. Austin's well-known theory of speech acts. However, the two theories were developed independently.

A speech act is a linguistic expression that not only explicates an assertion referring to an aspect of reality but has a "performative function"; that is, creates an act: "I hereby name this ship Queen Elisabeth." The main idea of this theory is that there are linguistic expressions, the nature of which is not descriptive. This is the same fundament by which Polanyi also differentiates between the two main types of linguistic expressions. According to this differentiation, the assertion "snow is white is true" is not just a description of reality but a tacit personal act; thus, if it is true that every linguistic expression that creates an act is a speech act, then "snow is white is true" is also a speech act. This conclusion, however, does not mean that "snow is white is true" is a speech act in the framework of Austin's theory, too, because Austin himself, contrary to Polanyi, regards true and similar phrases as linguistic expressions that are descriptive and simply refer to an aspect of reality.

This opposition shows that Polanyi tries not only to differentiate and to categorize speech acts and informative expressions but to argue that an explicit sentence in itself is not an assertion yet and that an explicit sentence becomes an assertion only by a tacit act of commitment. So, according to Polanyi, every expression is created by a tacit personal act, and certain linguistic expressions are not merely descriptive and explicate an assertion referring to an aspect of reality but create new performative acts at once. Therefore "snow is white is true," and similar expressions can only be regarded as speech acts if someone acknowledges that every expression is based on a tacit act of commitment; that is, in the case of these expressions of the latter type, verified at once by another tacit act—which is clearly not Austin's intention.

So, according to the logic of affirmation, meaning is not based on the explicit content of sentences but *on tacit acts of persons.* Affirmation is just as an act, like any other personal act. At the same time, due to its peculiar nature, affirmation gains a specific kind of meaning. Although Polanyi himself does not do this here, we can use his concepts of denotative and existential meaning to clarify this specificity (see ch. 4). Every tacit act of a person has existential meaning according to the person's commitment. However, an assertion of a sentence is a tacit act that has not only existential meaning but, because it refers to an aspect of reality, denotative meaning too. Traditional philosophy of language regards merely the denotative meaning as real. The reason for this is that only this kind of meaning can be connected to explicit sentences and thus impersonalized according to the objectivist ideal of knowledge.

Polanyi's arguments against the objectivist approach

In contrast to Polanyi, traditional philosophy of language tries to determine the meaning and usage of linguistic expressions in a formal, impersonalized way. Polanyi brings up several arguments against it.

We have seen that Polanyi differentiates between the confident and the skeptical use of a sentence. How can one decide that a person is using a word or a sentence confidently or skeptically? Polanyi claims that this can be done based *only on* the tacit component because exactly the same explicit content is used both in the confident assertion and the skeptical articulation of "snow is white." That is, only the tacit body language, tone, or context can help decide this question and not some kind of objective criteria. (Polanyi 1958, 250)

To evade this argument, the most typical solution to the problem is the so-called "ostensive definitions." Through these, seemingly objective connections can be made between certain linguistic expressions and certain aspects of reality, as somebody who knows the expressions correctly points to the referred objects.

> *"Accordingly, in formulating a definition we must rely on watching the way the art of using a word is authentically practiced; or more precisely, watch ourselves applying the term to be defined in ways that we regard as authentic. [...] The formalization of meaning relies therefore from the start on the practice of unformalized meaning."* (Polanyi 1958, 250)

This means that all ostensive definitions are based on the appropriate usage of language and the trust between the master and the learner; that is, on a kind of craftsmanship (see ch. 4). In other words, when somebody starts to learn a new language—by ostensive definitions, lexical explications, etc.—he uses that language and its expressions not skeptically but confidently, and he acknowledges that the given society uses that language properly. In this sense, his confidence in the meaning of the expressions of that language is *a tacit act of social loyalty.* So the determination of linguistic expressions by ostensive definition is not the least objective. (Polanyi 1958, 250-251)

Furthermore, Polanyi draws attention to the contradictory nature of the word "precise," but the following argument is also valid for other expressions, e.g., apt, fitting, clear, or expressive. According to Polanyi, "precise," similarly to "true," is not a denotative expression. For example, "this theorem is precise" means nothing other than the conclusion of the verification of that theorem asserts nothing new in terms of reality over the original theorem; thus, it cannot be verified in the same sense as the theorem referring to certain aspects of reality. Verification of a theorem due to the person's commitments is a tacit act concerning the assertion of the theorem and not just an explicit expression referring to some aspects of reality. Therefore, "precise," just as "true," is a tacit confirmation of an assertion and not a descriptive expression of an explicit sentence or a fact. (Polanyi 1958, 251-252)

On the other hand, Polanyi asks what the consequences would be if the expressions of language were perfectly exact, impersonal, and precise. The answer is the following: since reality is always changing, words will gradually lose their meaning because they would not be able to grasp the different, changing aspects of reality. Therefore, "only words of indeterminate meaning can have a bearing on reality [...]" (Polanyi 1958, 251). Furthermore, such perfectly formal and precise expressions would lose contact with persons too because:

> *"[...] any strictly formal operation would be impersonal and could not therefore convey the speaker's personal commitment."* (Polanyi 1958, 252)

As such, if, according to the objectivist approach, a person *consistently* took the criteria for perfect exactness and preciseness seriously, that would mean the final elimination of the person themselves.

> *"[...] any philosophy that sets up strictness of meaning as its ideal is self-contradictory. For if the active participation of the philosopher in meaning what he says is regarded by it as a defect which precludes the achievement of objective validity, it must reject itself by these standards."* (Polanyi 1958, 253-253)

This means that, according to the objectivist ideal of knowledge, the person and his tacit commitments are ignored in the structure of affirmation; that is, the meaning is restricted to the territory of explicit sentences. Therefore, only those aspects of reality can be acknowledged as real, which can be grasped by explicit expressions. But since the nature of a person and his commitments is tacit, if objectivism would be consistent, they cannot be regarded as real. This step, however, cannot be achieved because the person and his tacit commitments are the preconditions of every assertion. So objectivism is not consistent at all; it only builds a façade by deceptive substitutions (see ch. 1).

According to Polanyi, one has to examine not the impersonal sentences, which in themselves are meaningless, but the personal acts and tacit commitments by which someone asserts a sentence. *This examination has to be the aim of philosophy.* However, it clearly could not be objective since it would be about our most personal beliefs and convictions.

> *"Nothing that I shall say should claim the kind of objectivity to which in my belief no reasoning should ever aspire; namely that it proceeds by a strict process, the acceptance of which by the expositor, and his recommendation of which for acceptance by others, include no passionate impulse of his own."* (Polanyi 1958, 265)

The mind is not a logical inference machine

Since the revolution of information technologies of the forties and fifties, philosophers, psychologists, and neurologists, following the *computer-mind* analogy, generally try to model the workings of the human mind as a logical inference machine. Their goal, of course, is to describe them as objectively and exactly as possible.

According to Polanyi, the fundamental failure of the computer-mind analogy is that computers are merely logical inference machines. They do nothing other than transform the programmed, formally symbolized explicit sentences by strict formal rules. Since logical inference systems are clearly impersonal, they have *no* personal commitments or tacit beliefs concerning the explicit sentences and logical inferences at all. Consequently, according to the logic of affirmation, in their cases, there is neither assertion nor meaning, only the logical transformation of explicit sentences which in themselves without personal commitments are meaningless. The programmed, formally symbolized explicit sentences and the results of formal transformations are interpreted, in fact, by the *user* based on his own personal commitments and tacit beliefs. This is the reason that different users can interpret the exact same formal results so differently. If the user believes that he successfully eliminated the "unscientific subjectivity" of his person by the highly formal tools of computers, then he is committing a deceptive substitution. (Polanyi 1958, 258)

BOX 18: John Searle and the Chinese room

Polanyi's argument is strongly analogous to John Searle's famous Chinese room argument, which was explicated by the American philosopher against the concept of strong artificial intelligence. According to Searle, computers work exactly the same way and understand the real meaning of formal expressions, which they manipulate exactly to the same degree as the American man understands the Chinese expressions in the Chinese room.

Searle's thought experiment is the following: imagine a room in which there is an American man who knows nothing about the Chinese language. He has a large book with him that is a complex algorithm that tells him which Chinese phrases to use to answer specific Chinese phrases. Chinese people pass him sheets of paper through a window in the room, on which there are questions in Chinese. He, of course, does not understand a word, but he can find the answers in his big book. He writes them down on the sheets of paper and hands these back to the Chinese people who understand them and are happy with the answers.

It is obvious that the American man, thanks to his complex algorithm, is able to give appropriate answers in spite of the fact that he does not understand anything about the questions and answers. The reason for this fact is that he is not part of the Chinese linguistic environment in which the formal Chinese phrases gain meaning. The American man can only manipulate the formal linguistic phrases due to a logically determined mechanism. According to Searle, a computer works exactly the same way, and the formal manipulation of formal linguistic expressions cannot be regarded as real linguistic understanding.

Searle, of course, does not express his opinion in the framework of Polanyi's theory of personal knowledge, but, in terms of Polanyi's point of view, he definitely speaks about the difference between explicit sentences and tacit acts of understanding. Formal, explicit sentences in themselves remain meaningless if there is no tacit act of commitment by which a person can assert them. Computers, however, are not able to perform such tacit acts of commitment. They only manipulate formal, explicit sentences, but they do that much better than human beings. According to Polanyi, it is their real knowledge, not linguistic understanding.

Polanyi notes, referring to Gödel's incompleteness theorems, that even the purely formal systems are not perfectly exact and consistent; there is always some contradiction among the axioms. It follows, according to Polanyi, that "we never know altogether what our axioms mean since if we knew, we could avoid the possibility of asserting in one axiom what another axiom denies." (Polanyi 1958, 259) If one can make connections between the axioms and reality solely by asserting them, which, according to the logic of affirmation, includes a tacit component, then even the so formal and exact axioms cannot be perfectly explicit. (Polanyi 1958, 259)

Since only a person can give meaning to the symbol transformations of machines, they are subordinated to persons.

> *"Yet the necessary relatedness of machines to persons does essentially restrict the independence of a machine and reduce the status of automata in general below that of thinking persons."* (Polanyi 1958, 261-262)

This means that:

1. The user's mind defines the functions and goals of a machine.
2. The machine works in this framework.
3. The machine works under the control of the user's mind.

Polanyi claims that this structure is also valid for the situation of a neurologist examining and modeling the neurological processes of a human subject. (Polanyi 1958, 262)

1. The functions and goals of a neurological model are defined by the neurologist (based on the real or supposed workings and goals of the subject's—i.e., a human's—mind).
2. The neurological model works in this framework.
3. The neurological model works under the control of the neurologist and not under the mind of the subject.

According to the purely formal, explicit nature of the model, the subject's personal commitments and tacit beliefs, which give meaning to his acts and explicit sentences, are not represented at all in this model. It follows that the neurologist is not modeling the subject's mind and its real cognitive processes but merely the subsidiary parts of these processes; that is, the firings of neurons, the flows of ions along the dendrites, etc.

So at the end of the examination, the neurologist says *nothing* about the subject's mind, only about material conditions that entail the real cognitive processes of the subject's mind. The model becomes a legitimate description of reality only due to the fact that the neurologist denies the reality of every aspect of the process that is not represented by the model; i.e., he denies the reality of the subject's real cognitive acts. (Polanyi 1958, 263)

This situation leads to deceptive substitutions. A model based solely on logical inferences according to the computer-mind analogy does not represent the subject's commitments, beliefs, feelings, imaginations, desires, etc., all of which determine the subject's real cognitive processes, thoughts, and meanings. "Objectivism requires a specifiably functioning mindless knower." (Polanyi 1958, 264) However, the neurologist, as another human mind, knows well the real meanings of the workings of a human mind, and he tacitly, discreetly, substitutes these real meanings into his purely formal and explicit model. This is the only reason the neurologist could think that the model properly reflects the real cognitive processes of a human mind.

It follows, according to Polanyi, that the mind has to be acknowledged as an independent existent that cannot be reduced to explicit statements of material conditions.

> *"Our theory of knowledge is now seen to imply an ontology of the mind. [...] This notion—applied to man—implies in its turn a sociology in which the growth of thought is acknowledged as an independent force. And such a sociology is a declaration of loyalty to a society in which truth is respected and human thought is cultivated for its own sake."* (Polanyi 1958, 264)

However, he will establish this ontology only in ch. 13.

The fiduciary program

As we have seen with regard to the tacit nature of assertion, according to Polanyi, the main aim of philosophy should be the examination of personal commitments and tacit beliefs. The reason for this is that human persons have such deep evolutionary roots (millions of years) and cultural roots (thousands of years), which fundamentally determine their present, thinking, and goals. Strict skeptical methods, however, cannot examine the tacit acts that give meaning to explicit sentences; we can do that only in a confident way. Ignoring these tacit fundaments degrades assertions to vacant, meaningless sentences. It follows that nobody can start with a clean slate, and the fortress of philosophy cannot be constructed following the ideal of critical objectivity and formal preciseness because it would mean nothing. Self-reflectively, Polanyi applies this program to his *Personal Knowledge* at first. (Polanyi 1958, 264)

> *"I must admit now that I did not start the present reconsideration of my beliefs with a clean slate of unbelief. Far from it. I started as a person intellectually fashioned by a particular idiom, acquired through my affiliation to a civilization that prevailed in the places where I had grown up at this particular period of history. This has been the matrix of all my intellectual efforts. Within it I was to find my problem and seek the terms for its solution."* (Polanyi 1958, 252)

> *"Our tacit powers decide our adherence to a particular culture and sustain our intellectual, artistic, civic and religious deployment within its framework. The articulate life of man's mind is his specific contribution to the universe; by the convention of symbolic forms man has given birth and lasting existence to thought."* (Polanyi 1958, 264-265)

According to Polanyi, modern scientism based on the critical method "fetters thought as cruelly as ever the churches had done." (Polanyi 1958, 265) It follows that the rejection of objectivism and modern scientism, as well as the restoration of our trust in our personal commitments and beliefs, does not mean, of course, a reversal to the dogmatism of the Middle Ages.

> *"When I gave this book the sub-title 'Towards a Post-Critical Philosophy' I had this turning point in mind."* (Polanyi 1958, 265)

According to Polanyi, we have to accept that our deep evolutionary roots allow us to make contact with the hidden reality without any consistently objective criteria; that is, they allow us to formulate well-established, true scientific assertions about nature.

> *"The critical movement, which seems to be nearing the end of its course today, was perhaps the most fruitful effort ever sustained by the human mind. The past four or five centuries, which have gradually destroyed or overshadowed the whole medieval cosmos, have enriched us mentally and morally to an extent unrivalled by any period of similar duration. But its incandescence had fed on the combustion of the Christian heritage in the oxygen of Greek rationalism, and when this fuel was exhausted the critical framework itself burned away."* (Polanyi 1958, 265-266)

For Polanyi, the achievements of the Modern Ages are the *joint efforts* of the critical scientific movement (based on Greek rationalism) and the morality and belief system of Christian tradition, where the former is the driving force ("oxygen") at the forefront, and the latter is the tacit fundament hidden in the background, which was gradually sacrificed during the process. This should mean the end of the critical movement since it has no fundaments of its own upon which it could build further. Polanyi does not deny the efforts and achievements of the critical movement; on the contrary, he really appreciates them; however, he thinks that it became vain. He thinks it is the right moment to move on before the crisis—which also indicates the end of the critical era—significantly deepens. According to him, it is not just an epistemological crisis of science, but it affects the whole society—first and foremost its moral fundaments—and led to such symbolic events as the rise of Hitler and Stalin or the bombing of Hiroshima and Nagasaki (see ch. 7).

Polanyi refers to St. Augustine, for whom belief was the foundation of any knowledge. He contrasts him with John Locke, for whom belief was degraded into a mere lack of knowledge. This break in the seventeenth century was the reason that the critical mind rejected one of the main intellectual powers of the two. It is important to emphasize that this change was right and necessary in itself because the dogmatism of the Middle Ages represented a significant barrier before a free society and free thought. The crisis stems from the rejection of *all forms* of belief. Since our evolutionary roots, tacit skills, and intellectual passions work as beliefs, they compose the tacit fundament that allows us the scientific exploration of nature and the philosophical understanding of ourselves in it.

> *"We must now recognize belief once more as the source of all knowledge. Tacit assent and intellectual passions, the sharing of an idiom and of a cultural heritage, affiliation to a like-minded community: such are the impulses which shape our vision of the nature of things on which we rely for our mastery of things. No intelligence, however critical or original, can operate outside such a fiduciary framework."* (Polanyi 1958, 266)

It follows:

> *"[…] that the function of philosophic reflection consists in bringing to light, and affirming as my own, the beliefs implied in such of my thoughts and practices as I believe to be valid; that I must aim at discovering what I truly believe in and at formulating the convictions which I find myself holding."* (Polanyi 1958, 267)

Conclusion

Every articulate assertion has two composite parts: 1. *explicit sentence*; 2. *tacit act*. A sentence is always asserted by a person due to their tacit commitments, and this act gives meaning to the explicit sentence. Therefore, meaning is based on tacit acts of persons and not on explicit sentences, which only formulate the meaning.

According to Polanyi, during the Middle Ages, the significance of tacit beliefs and fundaments of knowledge was acknowledged. However, these fundaments and beliefs were forced into a false dogmatism because they were thought to be of divine origin and purely spiritual nature. Modern critical thinking put an end to this dogmatism, but since it regarded all real knowledge as objective and explicit and all tacit beliefs and convictions as a mere lack of knowledge, this process gradually led to the total rejection of all tacit commitments and beliefs and as a result, philosophy lost its original aim and real meaning. According to Polanyi, the mission of philosophy is to reveal our natural tacit beliefs and commitments and strengthen and deepen these tacit beliefs by explicit philosophical and scientific concepts and methods.

Chapter 9

The critique of doubt

Goals of the chapter

1. Polanyi tries to argue, by exposing the deficiencies and limits of critical philosophy, that there is a need for his post-critical philosophy.
2. He differentiates between the kinds of doubt in the following ways: tacit/implicit or explicit; contradictory or agnostic; juristic, religious, or scientific.
3. He shows that universal doubt cannot be consistent; that is, belief cannot be eliminated from knowing.
4. He uses historical examples to show that critical philosophy is just as based on "fiduciary" foundations; that is, based on uncritically accepted beliefs such as his own post-critical philosophy, and that, therefore, the claim that critical philosophy is more critical and thus more well-established it is only a facade. Moreover, since critical philosophy regards the critical method of doubt as to the foundation of knowledge, it conceals its own fundamental beliefs, becomes inconsistent, and does not execute even its own critical program. This execution, according to Polanyi, is not possible because if critical philosophy were to use its critical method on its own fundamental beliefs, it would eliminate itself. Furthermore, critical philosophy is also dangerous because, with the elimination of the tacitly accepted fundamental beliefs of science and society, the universe becomes meaningless, and disastrous consequences will follow, as already happened in the twentieth century.

***

BOX 19: Critical and post-critical philosophies

In the philosophical tradition, the term "critical philosophy" usually refers to the philosophy of Immanuel Kant and his followers. The term "critical," however, is used in a much broader sense, and critical thinking is opposed to "obscure," dogmatic thinking—although what the latter exactly means strongly depends on the given speaker. Furthermore, the concept of critical is tightly connected to the concept of skeptical.

Polanyi uses the concept of critical philosophy in a broad sense. He regards as critical every philosophy that tries to establish human knowledge and knowing on objective foundations and rejects belief as one of the two main conceptual skills of human beings. In his words: "The critical mind repudiated one of its two cognitive faculties." (Polanyi 1958, 266)

"*We must now recognize belief once more as the source of all knowledge. Tacit assent and intellectual passions, the sharing of an idiom and of a cultural heritage, affiliation to a like-minded community: such are the impulses which shape our vision of the nature of things on which we rely for our mastery of things. No intelligence, however critical or original, can operate outside such a fiduciary framework.*" (Polanyi 1958, 266)

According to this, critical philosophy was formed in the seventeenth century, and René Descartes and his methodological skepticism played the main role in the process. Polanyi calls his own philosophy post-critical as opposed to critical philosophy because he grounds it on a *fiduciary* fundament. Polanyi explicates the aim of *Personal Knowledge* as follows:

"*This book tries to serve a different and in a sense perhaps more ambitious purpose. Its aim is to re-equip men with the faculties which centuries of critical thought have taught them to distrust. The reader has been invited to use these faculties and contemplate thus a picture of things restored to their fairly obvious nature. This is all the book was meant to do.*" (Polanyi 1958, 381)

The true nature of the critical method

Polanyi explicates the point of the critical method as follows:

> "*The method of doubt is a logical corollary of objectivism. It trusts that the uprooting of all voluntary components of belief will leave behind unassailed a residue of knowledge that is completely determined by the objective evidence. Critical thought trusted this method unconditionally for avoiding error and establishing truth.*" (Polanyi 1958, 269)

According to the subjective-objective dichotomy of objectivism, *belief is subjective; knowledge is objective.* Therefore, such knowledge based on personal beliefs is also subjective; that is, not real knowledge at all. It follows that it has to be regarded by doubt and that only parts of it should be accepted as real knowledge that can be justified objectively; that is, that *can stand the tests of all kinds of doubt.*

BOX 20: Justified true belief

During the last centuries, the concept of "justified true belief" became the common definition of knowledge. Formally: S knows that a proposition P is true if and only if (1) P is true, (2) S believes that P is true, (3) S is justified in believing that P is true. According to this concept, knowledge is not belief but an essentially different thing due to a well-established verification by which it transcends belief.

Contrary to this, in Polanyi's opinion, belief is an intellectual skill by which a person can reach knowledge (see ch. 8). So for him, belief is not just an opinion that is true or not due to the actual evidence but the tacit fundament of any knowledge, including explicit scientific propositions.

The main goal of critical philosophy by the method of doubt is to overcome "medieval superstitions" and build a free, brave new world based on critical knowledge. This means, however, questioning traditional institutions; i.e., ultimately, the questioning of the principles of freedom—which are the foundations of modern societies—since these principles are not based on critical knowledge but gradually became parts of modern societies' fundamental beliefs and traditional institutions during the long course of cultural evolution. (Polanyi 1958, 269-270)

According to Polanyi, no one ever, in fact, wanted to execute the program of critical philosophy consistently, and it is not even possible to do so. (Polanyi 1958, 269-270) The reason for this is that every system of knowledge is based on *fundamental beliefs* or axioms that were accepted tacitly by trust, and this is also true even for such exact systems of knowledge as mathematics. Polanyi shows this fact through the discovery of non-Euclidian geometry. Therefore, since the fundamental beliefs or axioms of a system of knowledge are *preconditions* for any verification and argumentation in that system, they themselves cannot be verified and argued—not even by the so-called perfect method of doubt. (Polanyi 1958, 271)

Those who consider themselves strictly critical of everything merely do not know or ignore their own fundamental beliefs. This false self-image cannot be fixed as it was usually done in the name of self-reflection, by recognizing that there are certainly several serious mistakes in one's thinking regarding the critical method. The acknowledgment of this fallibility means that, according to strict, consistent objectivist criteria, one knows nothing. (Polanyi 1958, 271)

Polanyi claims that the roots of critical doubt go back not only to the sixteenth and seventeenth centuries but to human beings' primitive animal past. The everyday phenomenon of a person hesitating during a specific act about what to do next is referred to by Polanyi as implicit or tacit doubt. According to him, this deeper doubt is the foundation of explicit doubt in formal systems of knowledge, and to an extent, it determines every articulate assertion. If there were no tacit doubt, then there would have been no explicit doubt either—as it is the case with Laplace's demon, who never doubts anything although he is the ideal of critical thinking (see ch. 1 and 6). (Polanyi 1958, 272)

It does not follow for Polanyi that critical doubt has to be rejected in every case. He only argues that human thinking and systems of knowledge cannot be based on the critical method of doubt since the real fundaments of human thinking and knowledge are based on *trust.* As a matter of fact, Polanyi's harsh critique of the critical method is also a kind of doubt; only it is not directed against his own fundaments of thinking.

An example of when the method of doubt is useful is Polanyi's differentiation between contradictory and agnostic doubts. Every assertion can be explained in terms of doubt contradicting the originally given assertion. This is the case for contradictory doubt, which is fundamental in science. Suppose someone has a conviction: "I believe P," and they explicate it according to contradictory doubt: "I doubt P." Then, rephrase it into a positive affirmation: "I believe not-P" and show that not-P is not possible. By this method, the original conviction, "I believe P," can be confirmed even in the case of there being no direct evidence supporting it.

Polanyi claims that the fiduciary character of doubt is most obvious in the case of "reasonable doubt." It is clear that even the critical thinker himself does not want to doubt everything because that would lead to agnosticism and not to real knowledge based on doubt. Furthermore, dogmatic thinkers, of course, are also in doubt, only not in their own dogmas but in beliefs that cannot comply with their dogmas; for example, they doubt the fundamental beliefs of critical thinkers who believe that reasonable doubt is the source of all real knowledge. Since there is "dogmatic doubt" too, it is necessary to differentiate between reasonable and dogmatic doubts where the former is "well-grounded" and "systematic," and the latter is "unfounded" and "naïve."

> *"To urge that doubt must be reasonable, is to rely on something that cannot reasonably be doubted—that is, in legal phrase, a 'moral certainty.'"* (Polanyi 1958, 274)

This means that one suspends the critical method of doubt in the case of certain beliefs. However, this is not done based on the method itself, because the method, at least in theory, can be used against every belief, but it is based on something else outside the scope of the method, which, therefore, has to be an uncritical belief or conviction grounded on trust. In other words, since the critical thinker does not and cannot doubt everything, he relies on beliefs and convictions during his doubt in which he does not doubt, in which he trusts. In the positive sense, reasonable doubt can be separated from unreasonable doubt only by fundamental beliefs that are truer than the fundamental beliefs of unreasonable doubts.

> *"Natural scientists can be said to be more critical than astrologers only in so far as we regard their conception of stars and men as truer than that of the astrologers."* (Polanyi 1958, 274)

It follows that the method of doubt plays a positive role in science only in the sense that it can stabilize and defend the true beliefs and convictions of science against the attacks and doubts of other false systems of knowledge. The fundamental beliefs and truths of science, however, are not in the least the consequences of the critical method of doubt (see ch. 6).

Polanyi uses his formerly detailed example of Mesmer (see ch. 4) to show these limits of the critical method. The phenomenon of hypnosis was rejected by critical science *for almost a century* even though Mesmer and his followers had verified it with repeated and obvious experimental evidence. Today, this phenomenon is an integral part of critical science; for example, it is routinely used in different medical and psychological practices. Earlier, however, it went against the deep mechanical conviction about the nature of reality and, in the name of critical science, even the most obvious and palpable experimental facts about it were rejected.

The question of hypnosis has been solved. But since there have been several similar examples of critical science throughout the centuries, it can reasonably be supposed that there are also such phenomena today in science that are rejected because of the method of doubt, in the same way as for hypnosis. On the basis of these examples, it can be claimed that the method of doubt was executed poorly, and if, despite this fact, one does not, in turn, assert that today it can be used perfectly, then it follows that the method of doubt still necessarily leads to significant mistakes and unfounded scientific convictions. This means that the critical method of doubt in itself as a principle for scientific knowing is not established well; therefore, according to Polanyi, there is an opportunity for a post-critical philosophy. (Polanyi 1958, 274)

It follows that "the practice of scientific skepticism in respect to allegations rejected by science consists of upholding the current scientific view of their subject matter" and nothing more. Polanyi, however, reminds us that stability is a fundamental characteristic of science, and it worked well for centuries, and he, therefore, does not underestimate its importance. (Polanyi 1958, 276)

Polanyi claims that all scientific discoveries are conservative in the sense that they maintain and strengthen the scientific concepts and convictions on the basis of which they have been achieved. At the same time, there are *revolutionary events* that cause significant changes in the course of science; that is, in the explicit state of the framework of basic scientific concepts and convictions, because these events are heuristic and tacit in nature (see ch. 6). However, these changes do not create unbridgeable gaps in the development of science—which might appear to be the case according to strict objectivist criteria. Because these changes are *also* based on the scientific concepts and convictions that the conservative discoveries were, they merely grasp and unfold such parts and consequences of these former concepts and convictions that were not apparent to that point; that is, their meanings were tacit and hidden in the framework of the former concepts and convictions. (Polanyi 1958, 277)

Religion and doubt

Polanyi examines doubt in religion in detail because the critical method was originally used against Christian dogmas.

Polanyi claims that religion "is an indwelling rather than an affirmation." Therefore, its point is not the affirmation of the existence of God, as it was generally treated, but the immersion in a peculiar system of knowledge as an intellectual tool; that is, the *indwelling* itself. This process can be identified with the process that takes place when a swagger stick, as a tool, becomes a part of someone's hand—only it is realized at a higher existential level. (Polanyi 1958, 279)

> *"God cannot be observed, any more than truth or beauty can be observed. He exists in the sense that He is to be worshipped and obeyed, but not otherwise; not as a fact—any more than truth, beauty or justice exist as facts. All these, like God, are things which can be apprehended only in serving them."* (Polanyi 1958, 279)

This means that "God exists" is an affirmation of a tacit commitment, as the "snow is white is true" is an affirmation of the tacit act of claiming that "snow is white," and not an affirmation of an explicit fact referring to an aspect of reality as "snow is white" (see ch. 8). It follows that it cannot be doubted explicitly and merely rejected as a conviction or as truth. It can be doubted only tacitly, and in this sense, Christians always doubt it to some extent because this tacit doubt is an integral part of the Christian faith. (Polanyi 1958, 280-281)

So Christianity and the committed search for God cannot be disproved or understood by explicit facts. A Christian ceremony means nothing to someone who does not believe in God and is not a Christian. Nonetheless, there are factual testimonies in the history of Christianity that strengthen and confirm the Christian faith. During the last centuries, critical philosophy has successfully disproved a significant part of these testimonies. (Polanyi 1958, 283) However, since the worship of God is rather an indwelling, a commitment towards God, it cannot be disproved by disproving these factual testimonies or by any other scientific experiment. But it can be weakened. Factual events that did not happen, of course, have no meaning in relation to supernatural or any other systems of commitments. Biblical criticism is, therefore, successful not because of the refutation of the Christian faith but because of the refutation of these factual testimonies. (Polanyi 1958, 283-284)

> *"It destroyed the religious meaning of things without fully compensating for this loss by a different meaning, and the total volume of belief, from which all meaning flows, was effectively reduced."* (Polanyi 1958, 286)

Every affirmation is the consequence of a person's tacit commitment; explicit sentences in themselves mean nothing (ch. 8). In this sense, *belief is nothing else but a tacit commitment to reality that an aspect of reality has a peculiar nature, character, property, etc.* The roots of belief stem *from human evolution* and not from divine creation. Therefore, religious faith is a type of natural human belief when someone's tacit commitments are caged into a religious explicit conceptual and explanatory system. The role of religious traditions (among others) was to provide the intellectual tools by which human societies can explicate their fundamental tacit commitments. Of course, the consequence of this process was that what societies think about the universe, about their place within, and about their goals—that is, the meanings of the different aspects of reality—was determined by religion. According to Polanyi, we have to be thankful that critical philosophy destroyed religious dogmas and the static societies of the medieval and early modern eras and thereby made thinking freer. But we do not have to be thankful that it did not give answers to the most fundamental questions of human life and that, moreover, it often explicitly rejected these questions as unscientific. So we do not have to be thankful at all because, ultimately, the consequence of this process was that the universe has become meaningless.

> *"The book of Genesis and its great pictorial illustrations, like the frescoes of Michelangelo, remain a far more intelligent account of the nature and origin of the universe than the representation of the world as a chance collocation of atoms. For the biblical cosmology continues to express—however inadequately—the significance of the fact that the world exists and that man has emerged from it, while the scientific picture denies any meaning to the world, and indeed ignores all our most vital experience of this world."* (Polanyi 1958, 284-285)

However, the universe is not meaningless at all since the universe is home to life. Human beings have emerged from primordial material beginnings, and they not only give meaning to the world through their language but through their lives and existence, and, most of all, reveal and understand the inherent hidden rationality of reality (see ch. 1). In contrast to this, the objectivist-materialist approach merely offers a meaningless universe of deterministic-random material parts and, of course, those deceptive substitutions by which it conceals this meaningless picture of the universe from the public.

These are the necessary consequences of the critical method of doubt because: "All truth is but the external pole of belief, and to destroy all belief would be to deny all truth." (Polanyi 1958, 286) Truth cannot be asserted without belief. The affirmation of truth is a commitment to a certain meaning that an aspect of reality has a certain nature, process, character, etc.

Internal pole		*External pole*
Tacit commitment (belief)	→	**An aspect of reality (truth)**

Without belief and commitment, there remains but a set of meaningless facts of the parts.

> *"Objectivism has totally falsified our conception of truth, by exalting what we can know and prove, while covering up with ambiguous utterances all that we know and cannot prove, even though the latter knowledge underlies, and must ultimately set its seal to, all that we can prove."* (Polanyi 1958, 286)

Implicit beliefs

The most deeply rooted convictions of human nature are called implicit beliefs by Polanyi. These convictions are explained and determined by the conceptual frameworks of natural languages by which experiences are tacitly interpreted. The concepts and terms of assertions which are made about human experiences and different aspects of reality create the natural worldviews of human societies. (Polanyi 1958, 287)

> *"Our formally declared beliefs can be held to be true in the last resort only because of our logically anterior acceptance of a particular set of terms, from which all our references to reality are constructed."* (Polanyi 1958, 287)

Polanyi claims that personal commitments cannot be openly confessed today because they were caged into deep implicit levels by objectivism and the critical method of doubt. This was not the original goal of objectivism, of course. According to Polanyi, there is no explicit knowledge without tacit knowledge; that is, without improvable tacit convictions. These convictions cannot be made to cease to exist but can merely weaken and hide at deep implicit levels. (Polanyi 1958, 288)

Three aspects of stability

Implicit beliefs can be stabilized in three ways through strategies, intellectual tools that are used to defend the tacit beliefs and convictions of different conceptual and explanatory systems. (Polanyi 1958, 288-292)

1. *Circularity*: doubts that are kept off one by one strengthen the convictions against which they are explicated. Polanyi's example is the vocabulary of a language that involves a particular theory of the universe. The meanings of certain words that are temporarily under doubt are explained by other words that, of course, are not under doubt at that time. Then, these latter words are explained using words that were under doubt earlier. In the end, this process will strengthen the practice of the language and the trust in it.

2. *Epicyclisity*: this means the gradual expansion of a conceptual and explanatory system by new auxiliary theories that can explain the contradictions and solve the problematic points of the system. So the convictions that have been attacked by doubt are strengthened by new auxiliary theories, and then if these new theories are also attacked, the same auxiliary process will help these theories too.

3. *The principle of suppressed nucleation*: the experiences that could be evidence for other rival conceptual systems are examined and interpreted one by one; thus, they can be refuted one by one, which will prevent the formation of any possible rival systems. A given experience in itself can always be explained, for example, by a new auxiliary theory. So, even if a new conceptual system has greater explanatory power than the old one, if it cannot function as a comprehensive theory, its lone parts would be refuted successfully.

These strategies, however, assert nothing about the truth of the defended conceptual and explanatory systems. That is another question that depends on the relation between reality and the conceptual systems. (Polanyi 1958, 292)

The objectivist conceptual system and modern science secure their stability also by these strategies, as does most superstitious witchcraft. (Polanyi 1958, 292) See, e.g., the treatment of anomalies. (Polanyi 1958, 293) However, according to Polanyi, it is perfectly fitting if a true conceptual system has to defend its truth. (Polanyi 1958, 294)

Polanyi regards coherence as the criteria of stability and not of truth since coherence is but a formal consistency and not a fiduciary act, a commitment to a given system of knowledge. So coherence can strengthen even such a system that is false to the core. According to Polanyi:

> *"[...] there exists no principle of doubt the operation of which will discover for us which of two systems of implicit beliefs is true—except in the sense that we will admit decisive evidence against the one we do not believe to be true, and not against the other. Once more, the admission of doubt proves here to be as clearly an act of belief as does the non-admission of doubt."* (Polanyi 1958, 294)

The fiduciary basis of rationality

At the end of the chapter, Polanyi emphasizes again that universal doubt, according to the true nature of the Laplacian ideal of knowledge, destroys all meaning because no one can doubt his every concept and belief at the same time since doing so would mean that none remain by which he could reinterpret them. (Polanyi 1958, 294) So rationality is not based on pure doubt but on fiduciary roots. (Polanyi 1958, 297)

> *"What they [the critical thinkers] actually want is not expressed but concealed by their declared principles. They want their own beliefs to be taught to children and accepted by everybody, for they are convinced that this would save the world from error and strife."* (Polanyi 1958, 297)

> *"Since the skeptic does not consider it rational to doubt what he himself believes, the advocacy of 'rational doubt' is merely the skeptic's way of advocating his own beliefs."* (Polanyi 1958, 297)

Polanyi cites Bertrand Russell: "Thus rational doubt alone, if it could be generated, would suffice to introduce the Millennium." That is, the upcoming brave new world. According to Polanyi, the real meaning of this prophecy should be explicated by the following: "The acceptance of rational beliefs such as my own would suffice to introduce the Millennium." (Polanyi 1958, 297) But based on this honest wording, modern scientism would have to give up the illusory principle of doubt, which would lead to a disdained situation where critical philosophy would have to face its own fiduciary fundaments. (Polanyi 1958, 298)

According to Polanyi, the early rationalist of the seventeenth and eighteenth centuries stood up against early modern religious dogmatism and rightfully used the critical method of doubt. Fortunately, in that era, this method did not yet attack their tacit beliefs and convictions. At the beginning of the twentieth century, however, the method of doubt became totally useless against the new revolutionary dogmatisms, for example, mainly against Marxism and National Socialism. By that time, the critical method had already entirely undermined the fundamental beliefs and convictions of modern rationalism. Modern rationalism and philosophy can stand up against these revolutionary menaces only through their true beliefs and traditions of the eighteenth and nineteenth centuries because these revolutionary systems use the exact critical method of doubt against rationality and science. So if modern rationalism, science, and philosophy still try to use the critical method against these menaces, this will undermine even more the true beliefs of rationalism and modern societies and strengthen these revolutionary systems.

> *"Modern fanaticism is rooted in an extreme skepticism which can only be strengthened, not shaken, by further doses of universal doubt."* (Polanyi 1958, 298)

One can stand up against fanaticism merely on the basis of trust. (Polanyi 1958, 298)

Conclusion

The critical method of doubt is not the real foundation of knowledge. Every conceptual and explanatory system has its own fiduciary basis that was never doubted. Since if someone doubted everything, even their own fundamental beliefs, then she could consistently assert nothing—but no one tries to follow this consistent path of the critical method.

The method of doubt stems from the nature of objectivism, which, according to the false subjective-objective dichotomy, regards every belief as subjective and merely accepts those statements as knowledge that has been proved by some impersonal explicit process. Objectivism, however, ignores the fact that, without the prior acknowledgment of certain fundamental principles or axioms, no process of verification or argumentation can be accomplished; that is, certain fundamental principles *per definitionem* have to be acknowledged on the basis of trust, even in the case of the most exact sciences and mathematics.

The method of doubt was originally focused on medieval and early modern religious dogmas; therefore, its role was not to establish the fundamental principles of philosophy and scientific knowledge at all but to stabilize and defend them. The rejection of religious dogmas and the liberation of thinking were wonderful achievements of the modern era. Belief, however, cannot be identified with religious dogmas because it is a natural human commitment towards reality, by which any truth at all can be affirmed. The explicit rejection of every kind of belief due to the critical method of doubt necessarily leads to the weakening and concealing of the real fiduciary principles of knowledge. This process poisoned twentieth-century Europe, and it can be halted only if philosophy and science set an example to the whole Western world and honestly return to their real tacit fiduciary foundations; that is, a post-critical era starts.

Chapter 10

Commitment

Goals of the chapter

1. According to the logic of affirmation, a fiduciary act is a necessary part of any affirmation (see ch. 8). In this chapter, Polanyi shows his approach, by which he reflects on his own fallibility. The main point of this approach is that, in contrast to the method of doubt, it is coherent and not self-destructive.

2. The solution to the problem is the *act of commitment.* Without any fundament of knowledge that complies with the objectivist ideal, it grounds the responsible affirmation of truth and the use of cyclical conceptual systems.

3. Polanyi presents his concept of personal knowledge. Its fundament is a commitment he calls *universal intention,* which makes contact between a person and reality. Since this reality is impersonal, it is independent of the person.

4. Polanyi establishes the separation of personal from subjective. According to him, subjective states have no universal intention, which means that the person does not act in these cases; she is only the subject of these states, and therefore these states lack any commitment.

5. Polanyi shows the rightful place of passions in knowing and scientific discovery.

6. He shows the consequences of the critical method of doubt with the help of the concept of "objectivist dilemma."

   "*[...] commitment is a personal choice, seeking, and eventually accepting, something believed [...] to be impersonally given, while the subjective is altogether in the nature of a condition to which the person in question is subject.*" (Polanyi 1958, 302)

7. Polanyi defines the relation between his theory and the relativist approach. He argues that, although a personal act of commitment is a precondition of knowledge, there is only one truth. Furthermore, although it is true that the commitments of a person are influenced by their society, this impact is neither coercive nor arbitrary in direction and degree.

In the fifties, when *Personal Knowledge* was written, relativism was not a central issue. Thus, Polanyi argues mainly against objectivism and defines his philosophical approach as contrary to it. He makes clear several times, however, about his approach: "Nor is it relativistic." (Polanyi 1958, 316) It follows that Polanyi, in fact, defines his theory of personal knowledge in contrast to both objectivism and relativism.

***

The nature of fundamental beliefs

As we have seen in the previous chapter, the objectivist approach wants to build knowledge on a fundament that cannot be doubted; that is, it has stood the proof of any kind of critic. According to Polanyi, it is not possible, and, reflecting on this fact, he acknowledges that his own approach of personal knowledge is necessarily *circular.*

> *"Any enquiry into our ultimate beliefs can be consistent only if it presupposes its own conclusions. It must be intentionally circular."* (Polanyi 1958, 299)

This circularity could only be broken if an approach to knowledge was possible that could question its own fundamental beliefs before it would even accept them to start on further inquiries based on these fundaments. This would mean that (a), it is possible to reach knowledge without any fundaments, or (b), there are purely rational fundamental beliefs (in this case rather axioms) that cannot be questioned—that is, that *a priori* can be acknowledged; so there is no need for further examination in their case. Polanyi claims that neither of these cases is consistent. According to him, an honest philosophy has to face the fact that it has fundamental beliefs that were acknowledged without any critical examination based on an *acritical fiduciary* act; and that it can start on further inquiries only after this fiduciary act when it has fundaments upon which it can build any new pieces of knowledge.

In the case of the examination of fundamental beliefs, this means the presupposition of the conclusions; i.e., the examination of fundamental beliefs starts from the fundamental beliefs that have been acknowledged by a fiduciary act as the basis for any examination. In the end, the examination will arrive at the fundamental beliefs, which will have been strengthened and deepened by the examination; otherwise, it would refute its own fundaments and thus itself. "*Logically, the whole of my argument is but an elaboration of this circle.*" (Polanyi 1958, 299) This is the reason that he mentions:

"I cannot hope to do more in this book than to exhibit a possibility which like-minded people may wish to explore." (Polanyi 1958, 315)

In a logical sense, Polanyi does not try to use coercive arguments because he denies the existence of such arguments. He can only hope that like-minded persons who accept the same fundamental beliefs will find his honest and consistent alternative approach to knowledge persuasive. (Polanyi 1958, 315)

The main fundamental beliefs of science and philosophy

Polanyi commits himself to the following fundamental belief:

"I believe that in spite of the hazards involved, I am called upon to search for the truth and state my findings." (Polanyi 1958, 299)

The hazards stem from the fact that his fundamental beliefs have been acknowledged by a fiduciary act and not by an unquestionable critical examination, so, in principle, it can turn out at any time that they are false or misinterpreted.

Perhaps the most important fundamental beliefs of a scientist are *to search for the truth* and *to share the truth with others.* The search for truth is a heuristic act since there are no critically unquestionable fundamental beliefs from which any truth can be deduced. The act of persuasion is also heuristic in nature because there are no logically coercive arguments according to which everybody has to accept the newly discovered truth. (Polanyi 1958, 299)

The method for reaching the truth

According to the objectivist ideal of knowledge, only knowledge that has stood the proof of the critical method of doubt can be regarded as the truth, i.e., be unquestionable. As we have seen in the previous chapters, objectivism is not consistent at all; thus, in practice, it does not even try to follow its own ideals (see ch. 1). Contrary to objectivism, Polanyi acknowledges that there is no unquestionable knowledge; that is, he accepts a concept of knowledge that is consistent and complies with scientific practice. Therefore, in contrast to cold, critical examination, the discovery of truth, in reality, happens in the following way:

"The personal participation of the knower in the knowledge he believes himself to possess takes place within a flow of passion. We recognize intellectual beauty as a guide to discovery and as a mark of truth." (Polanyi 1958, 300)

This means that the search for the truth is not based on any logically coercive objective method but on intellectual passions. (Polanyi 1958, 300) In other words: truth cannot simply be deduced by an explicit method from objective (focal) facts; truth can merely be recognized, revealed, and then acknowledged by tacit skills and intellectual passions. (Polanyi 1958, 300)

The two poles of personal knowledge

Polanyi differentiates between *personal* and *subjective* passions. A person merely endures his subjective passions and feelings. These passions are passive because they are not parts of those commitments that go beyond the person. However, the main characteristic of personal passions is precisely the fact that their goals are far beyond the person, and they actively encourage him, for example, to search for the truth. Therefore, the personal "transcends the disjunction between subjective and objective." (Polanyi 1958, 300)

A personal act is a kind of act that, on the one hand, is motivated and guided by individual passions. Thus, it cannot be objective and, on the other hand, is submitted to impersonal principles that are independent of the person, such that, due to these principles, it cannot be merely subjective. It follows that personal knowledge has two poles, a *universal* and a *subjective* one, which are connected by the act of commitment. (Polanyi 1958, 300)

The two poles of personal knowledge:

Commitment

Universal ←——— **Passionate person**

The answer to the question of why the nature and structure of personal knowledge are so can be found in the evolutionary origin of human beings. According to this, Polanyi defines the territory of personal in human intellectual life:

1. At the lower end of this scale lie perception and the satisfaction of appetites, which are mainly controlled by unconscious personal judgments. (In the case of perception, however, it happens from time to time that the automatic, passive processes are overruled by conscious intellectual acts.) Bodily processes under the intellectual level are, of course, tightly connected to the person himself; however, these processes do not commit the person because they can be corrected, as happens quite often. This means that the person does not entirely identify himself with the inarticulate

person of the body to whom these lower-level desires and motivations belong. Therefore, the participation of the person in these processes is severely limited; it is, rather, the territory of the subjective. (Polanyi 1958, 301)

2. For other reasons, the person's participation is also restricted in the higher-level intellectual processes. These are the territories of mathematics and partly of exact sciences, where the intellectual processes are controlled by strict formal rules and schemes that are, in a sense, maiming all real personal activities and judgments. Since this kind of thinking is at the higher level of the intellectual scale, it can be regarded as natural that the objectivist ideal of entirely exact and formal knowledge was born. (Polanyi 1958, 301)

3. The real territory of personal lies between these two endpoints. It is the territory of articulate problems that can only be solved, on the one hand, by tacit heuristic passions and, on the other, in a framework of a coherent, explicit intellectual scheme. (Polanyi 1958, 301) Polanyi claims that the best example of this is scientific discovery itself:

   "*From the first intimation of a hidden problem and throughout its pursuit to the point of its solution, the process of discovery is guided by a personal vision and sustained by a personal conviction.*" (Polanyi 1958, 301)

This means that, although thinking goes forward in an explicit intellectual framework, it has to overcome the logical gaps by tacit heuristic and passionate efforts, and if it is necessary, it has to reinterpret or modify the intellectual framework itself radically. According to this:

> "*While originality conflicts sharply with the ideal of a completely formalized intelligence, it also differs altogether from drive-satisfaction. For our appetites are ours, and it is ourselves they seek to satisfy, while the discoverer seeks a solution to a problem that is satisfying and compelling both for himself and everybody else. Discovery is an act in which satisfaction, submission and universal legislation are indissolubly combined.*" (Polanyi 1958, 301)

So the satisfaction of a desire and the universal intention towards a principle are the two endpoints of personal knowledge. However, universal principles can exist to someone only if the given person commits himself to them and submits his own personal acts and judgments to these universal principles.

Without these acknowledged and followed universal principles, there remains merely the subjective satisfaction of desires. In other words, if there are no universal principles, knowledge necessarily remains subjective, which leads to relativism. Universal principles, however, do not exist as facts but *as the truth.* They become valid solely by a personal act if someone commits himself to them. "Commitment is in this sense the only path for approaching the universally valid." (Polanyi 1958, 303)

This means that universal principles cannot be grasped in the same way as can be done with tangible things thanks to a simple knowing process; for example, "snow is white" can be described by an explicit sentence and can then be asserted. Universal principles can be grasped in a way that reflects the structure of the affirmation of "snow is white is true"; that is, universal principles can only be grasped by a *tacit act,* by a *commitment* as a person commits himself to those principles (see ch. 8 and 9). Without this, they will lose their meaning. (Polanyi 1958, 308)

The objectivist dilemma

The conflict that necessarily persists between the objectivist requirement of impersonality and the natural human aspiration that urges a person to commit himself to the truth is referred to by Polanyi as the *objectivist dilemma.* (Polanyi 1958, 304) The former is the objectivist precondition of truth since, according to objectivism, to avoid subjectivity, an assertion can be true if and only if it is entirely impersonal. At the same time, without the latter—that is, without a personal commitment—there is no truth, only so-called "objective facts." An example of this is the sentence "snow is white," which means nothing in itself if a person does not assert it by a personal act of commitment, saying, "I believe that snow is white." However, this personal commitment does not correspond to the objectivist criteria of truth. From a consistent view of impersonality, all personal passion is simply a false subjective belief. Thus, all assertion by personal commitment is a meaningless sentence, and believing all facts to be true on this personal basis is a vacant supposition because there is no connection between a subjective commitment and objective reality. This means that, from a consistent view of impersonality, nothing can be true. (Polanyi 1958, 303) (This situation complies again with the meaningless "knowledge" of Laplace's demon; see ch. 1 and 6.)

> *"For it is self-contradictory to secede from the commitment situation as regards the beliefs held within it, but to remain committed to the same beliefs in acknowledging their factual content as true."* (Polanyi 1958, 304)

For example, the assertion that "I believe that snow is white" emphasizes the personal pole according to the nature of heuristic activities, and "snow is white is true" emphasizes the universal pole, which generally arises in the act of persuasion. But the meanings of the two assertions are essentially the same; that is, a person has committed himself to a truth that "snow is white," which is independent of him and universally valid. (Polanyi 1958, 305)

> *"According to the logic of commitment, truth is something that can be thought of only by believing it."* (Polanyi 1958, 305)

Responsible free act

The precondition that enforces a responsible free act is *a person's need for universality.* While a person acts according to his desires, he is free only in the sense that he follows his *own* desires. Thus, he is not restricted by external forces and does not have to be responsible for anybody or anything in his choices. However, the need for universality acknowledged by a person urges him to constructive efforts that will coerce his acts: according to the universal intention of his commitment, he will freely act on his own will as he *has to* act. (Polanyi 1958, 309) "Here I stand and cannot otherwise," Polanyi cites Luther. (Polanyi 1958, 308)

> *"The freedom of the subjective person to do as he pleases is overruled by the freedom of the responsible person to act as he must."* (Polanyi 1958, 309)

So if a person acts conscientiously according to his responsible situation, although he has no choices, he still freely chooses what he has to choose. In this case, the person is not restricted by external forces, just as is the case in acts by subjective desires, since the person's choice stems from his *own commitment* and his *own sense of responsibility.* From another point of view, there are no free choices in this latter case either, because the person's actual subjective desires determine his choice entirely. The main difference is in the intentions of the two kinds of acts and the natures of the two kinds of determinations. In the case of subjective desires, a person's intention is solely towards satisfying his own bodily desires, while in the case of a responsible free person, his acts are determined by the universal intention of his commitment, which goes beyond the person towards a *universal principle.* This is the reason that the former person is free only in a very limited, subjective sense, while the latter is free in the universal sense.

The act of a responsible free person is always heuristic since there is an inevitable indeterminacy, a logical gap in his acts; thus, the responsible person has to bridge these gaps by his own active heuristic efforts (Polanyi's example is scientific discovery). This significantly differs from the nature of the passive act/determination of desires. The acts of responsible persons become determined by the fact that "they are submitting to the universal status of the hidden reality which they are trying to approach"; that is, their acts are determined merely by a hidden reality they want to discover. (Polanyi 1958, 309) (This difference is, in fact, between a goal-directed and an effect-like determination which Polanyi calls "*a fronte*" and "*a tergo*" determinations. He details them in ch. 13.)

Merely the act of commitment, therefore, establishes a relation with reality. (Polanyi 1958, 304) If a person says a sentence in relation to reality without any commitment—for example, "it is night"—then the content of the sentence can comply with reality only in that random situation if it is, in fact, night. However, "Responsible action excludes randomness, even as it suppresses egocentric arbitrariness." (Polanyi 1958, 310) Of course, random events impact any responsible free acts, but they do not guide them. (Polanyi 1958, 309)

Commitment and the truth of scientific discovery

The goal of problem-solving or scientific discovery is but the revelation of the hidden reality. These processes are heuristic in nature. It follows that they cannot be mechanistic and logically coercive since the continuously changing nature of reality excludes these possibilities (see in ch. 13).

> *"The implications of new knowledge can never be known at its birth. For it speaks of something real, and to attribute reality to something is to express the belief that its presence will yet show up in an indefinite number of unpredictable ways.*
> *An empirical statement is true to the extent to which it reveals an aspect of reality, a reality largely hidden to us, and existing therefore independently of our knowing it. By trying to say something that is true about a reality believed to be existing independently of our knowing it, all assertions of fact necessarily carry* universal intent. Our claim to speak of reality serves thus as the external anchoring of our commitment in making a factual statement." (Polanyi 1958, 311)

If the major part of reality is not known, and this is a fact, then it means that these hidden parts of reality exist *independently* from human knowing and manifest themselves in *unpredictable* ways. And if assertions of facts refer to such aspects of reality that originally exist independently from human knowing, then it means that these assertions of facts are universally intended and are true; therefore, everybody has to accept them as true facts. It follows:

> *"In a heuristic commitment, affirmation, surrender and legislation are fused into a single thought, bearing on a hidden reality."* (Polanyi 1958, 311)

> *"A result obtained by applying strict rules mechanically, without committing anyone personally, can mean nothing to anybody."* (Polanyi 1958, 311)

This last quotation again refers to the ideal "knowledge" of Laplace's demon (see ch 1 and 6). This "knowledge" is the consequence of perfect and logically coercive rules (the ideal equations of physics) but means nothing because Laplace's demon is a bodiless, purely rational spirit who, *per definitionem, cannot commit himself* to anything. He "knows" all the possible explicit sentences concerning reality, but without tacit skills, commitments, and universal intentions, he cannot assert these sentences to reality. Therefore, the objectivist ideal of knowledge, the Laplacian super intelligence, in fact, knows nothing. The system of perfect and exact explicit sentences "known" by it means anything merely to human persons. It seems that Laplace's demon owns this knowledge only because of a deceptive substitution.

It follows, according to Polanyi, that scientific discovery is not a logically coercive mechanistic process but the *highest-level craftsmanship*. It is the highest level because it uses the most improved tangible and intellectual tools; however, its point is the same as that of the simplest craftsmanship; it is based on tacit human skills. (Polanyi 1958, 311-312) As with every other intentional act, scientific discovery is also based on unconscious bodily functions and is also limited by inborn and gradually developed skills of childhood. Moreover:

> *"Worse still, we are creatures of circumstance. [...] And this reliance on the cultural machinery of our society continues through life."* (Polanyi 1958, 322)

So scientific discovery is an active heuristic act, and since it is not logically coercive, it always includes the possibility of failure. It follows that human beings can believe only in things that could be false. (Polanyi 1958, 313) However, the fact that human beings can be wrong does not mean that there is more than one truth according to the different convictions of different persons. In Polanyi's words:

> *"Therefore, though every person may believe something different to be true, there is only one truth."* (Polanyi 1958, 315)

Assertions referring to truth are, of course, influenced by "the exigencies of human intercourse," historical eventualities, the consensus of the given community, etc. But, according to the nature of universal intentions, the point of a scientific assertion is always that need for truth on the basis of which the assertion is formulated and the need that this assertion has to become the ruling consensus. As we have seen in ch. 8, there is no essential difference between an assertion and the assertion of the truth of an assertion. The latter is but the confirmation and emphasis of the universal pole of the former assertion according to the act of persuasion. It is entirely different from this to speak about what the consensus in a given community is or what someone else thinks, or even what the person thought in the past since *these statements do not refer to the truth but to assertions of others.* It follows that, during these statements, the person does not commit himself to the content of these assertions at all. Truth has simply nothing to do with these statements. (Polanyi 1958, 316)

"Truth" of other conceptual systems; relativism and objectivism

A question arises: if there is only one truth, then what does a person have to think about the "truth" of other conceptual systems, which is different than the truth in which the person believes according to his conceptual system and convictions? Polanyi's answer is simple: he has to reject it as false a belief. A fiduciary (acritical) philosophy does not eliminate doubt itself; it merely denies that the method of doubt is the fundamental method for searching the truth. That is, in contrast to critical philosophy, it claims that there are fundamental beliefs and convictions that a person, even in that case, has to insist on when he could doubt them. (Polanyi 1958, 318)

According to Polanyi, a conceptual and explanatory system with another "truth" than ours can be rational and verified due to its own methods and principles. Critical philosophy, however, has to regard such conceptual systems as irrational because they are not based on the "reasonable doubt" of critical philosophy (see ch. 9). Nevertheless, despite the fact that, due to its own methods and principles, every conceptual system can be rational and verified, it does not follow that any given conceptual system can be true as relativism holds. Polanyi refers to these conceptual systems, which are rational due to their own methods, as *competent.* But in spite of their competency, they can be false. Competency and truth can replace each other merely by deceptive substitutions.

Polanyi differentiates between competent and purely subjective conceptual systems, the latter of which are entirely illusory. This important distinction was denied by both objectivism and relativism, because the former claims that there is only "one real" competency by which truth can be reached, which is "reasonable doubt," of course, and the latter acknowledges as "truth" the results of every kind of competency. For objectivism, therefore, every other conceptual system is incompetent and false, and for relativism, every one of them is competent and "true," while Polanyi claims that several of them are incompetent and false, and the others are competent and false.

It follows that there are logical gaps between conceptual systems that hold themselves to different methods and principles, as the degree of competency and these logical gaps cannot be bridged by the explicit methods of either conceptual system. The main consequence of this fact is that they have to struggle with each other for survival. Although human persons from different conceptual systems can be—and, for different (moral) reasons, have to be—tolerant with each other, the different conceptual systems necessarily question each other's intellectual existence because all of them claim the truth for themselves with universal intention, and there is only one truth. (Polanyi 1958, 318)

So Polanyi, similarly to relativism, acknowledges that other conceptual systems can be rational and competent; moreover, he acknowledges that environment, upbringing, culture, the ruling consensus, etc., all severely impact the tacit skills, methods, and principles by which a person searches the truth. But if this is the case, then on what basis can somebody claim that he has the truth? The answer is the following: the fact that environment, upbringing, culture, etc., severely impact a person leads to the concept that all personal belief is reduced to "the mere products of a particular location and interest" based merely on critical philosophy. (Polanyi 1958, 322) But Polanyi himself says:

> *"I accept these accidents of personal existence as the concrete opportunities for exercising our personal responsibility. This acceptance is the sense of my calling." (Polanyi 1958, 322)*

According to critical philosophy, therefore, these external factors as subjective factors obstruct the critical method, and according to relativism, they absolutely determine any personal beliefs and convictions; however, Polanyi claims that these factors are only the *conditions* and also the actual *opportunities* of the situation in which a responsible person has to act to reach the truth by his own heuristic efforts. This means that these factors can obstruct heuristic efforts and create conditions and opportunities by which a person can successfully fulfill his calling. However, in its essence, success—that is, to reach and reveal the hidden reality and to assert the truth—is *always the achievement of personal efforts.* It is our responsibility and the consequence of those external factors into which we are born. (Polanyi 1958, 322)

> *"I believe, therefore, that as I am called upon to live and die in this body, struggling to satisfy its desires, recording my impressions by aid of such sense organs as it is equipped with, and acting through the puny machinery of my brain, my nerves and my muscles, so I am called upon also to acquire the instruments of intelligence from my early surroundings and to use these particular instruments to fulfill the universal obligations to which I am subject."* (Polanyi 1958, 323)

According to Polanyi's post-critical philosophy, a person has to accept his personal circumstances as *opportunity-conditions* because he cannot do it any other way. If according to objectivism, he tries to reject them as subjective obstacles, then he cannot even consistently speak his truth because every language is an integral part of the random factors of opportunity-conditions in which he has to get along. (Polanyi 1958, 314-315) And if according to relativism, he acknowledges that these external factors absolutely determine his beliefs, then he should consistently disclaim every kind of truth.

Conclusion

According to Polanyi, the main fundamental belief of a scientist is their commitment to truth. Polanyi agrees with objectivism that there is truth and universally valid criteria for knowledge. Polanyi is not a relativist. At the same time, he claims that neither truth nor any universally valid criteria can be formulated in a purely explicit, objective way, but that they can merely be grasped by a tacit commitment of universal intentions. It follows that personal knowledge has two poles: one personal, which anchors every human person to a peculiar biological body, to a language, to a culture; and a universal one, which goes beyond the person and defines the universally valid criteria and truths to which he submits himself to act.

Part Four: Knowing and being

Chapter 11

The logic of achievement

Goals of the chapter

1. Polanyi shows that neither machines nor living beings and human thinking can be explained based only on physical, mechanical knowledge.
2. The reason for this is that the workings and actions of these comprehensive entities include achievements, the appraisal of which can be based only on personal knowledge.
3. In a general sense, this required knowledge refers to different *rules of rightness*, which Polanyi calls *operational principles* in the case of machines. These operational principles define the successful workings of machines, and they cannot be acknowledged and explained based only on physical-chemical descriptions.
4. Polanyi shows that an activity or working can be regarded as erroneous only if the rules of rightness of that activity or working have been established. Physical-chemical knowledge is required for the explanation of a malfunction because only this kind of knowledge can show in which specific physical-chemical conditions a machine's operational principles can function.
5. Polanyi claims that this type of relationship between machines and their physical-chemical parts can be found between the different achievements of living beings and their bodily parts. He exemplifies this with the relation between logical thinking and psychology: logical thinking adheres to definite rules of rightness, but it requires the subsidiary presence of certain appropriate physiological states and faculties. Psychology in itself cannot explain logical thinking; however, it can explain the inferential errors compared to the rules of rightness in question.

***

Epistemology

Polanyi's argument with regards to epistemology is that the concept of tacit and personal knowledge—contrary to mainstream critical epistemologies—*acknowledges* and, due to its specific structure, has to acknowledge the comprehensive achievements of machines and living beings that cannot be deduced from its composite physical-chemical parts. According to Polanyi, the particular natures of machines, living beings, and human thinking, which cannot be reduced to their composite parts, *tacitly* are acknowledged by everybody since it concerns our natural ontological convictions towards them. However, he details and argues this claim only in ch. 13.

> *"Our comprehension of a living individual entails a subsidiary awareness of its parts which is not wholly specifiable in more detached terms. This understanding acknowledges a particular comprehensive—i.e. 'molar': achievement of the individual itself. Since our knowledge of this molar function is not specifiable in 'molecular' terms, the function itself is not reducible to molecular particulars; it must be acknowledged therefore as a higher form of being, not determined by these particulars."* (Polanyi 1958, 327)

Here Polanyi relies on those earlier results (see mainly in ch. 4), according to which, a recognition of a comprehensive ("molar") whole cannot be deduced from the ("molecular") knowledge of the parts which, during the recognition of the whole, we are aware of only subsidiarily. This process is, in fact, an achievement—more precisely, an achievement of recognition—that is, it is based on personal knowledge. The subsidiary parts of the whole can be recognized and explained only through the recognition of the whole; that is, only *after* the recognition of the whole. In itself, the objective, Laplacian knowledge of the (focal) parts—which in this case, in fact, are not parts of anything—does not and cannot include the comprehensive knowledge of the whole referring to its functions, workings, operational principles, etc. (Polanyi 1958, 327) Polanyi calls the latter, which cannot be determined without personal participation and personal knowledge, *personal facts.* (Polanyi 1958, 343)

It follows that the higher-level knowledge referring to the whole cannot be reduced to the lower-level knowledge referring to the parts; that is, *the reality* of the higher-level comprehensive whole—and the reality of its functions and achievements—*has to be acknowledged.*

> *"Of course, every comprehension of a whole acknowledges the reality of it; and whatever we comprehend both means something to us and to a certain extent at least means something also existentially, in itself."* (Polanyi 1958, 344)

This is the argument in the most general sense that will be repeated in this and the next two chapters in different versions. (Polanyi 1958, 327) However, Polanyi will establish the emergent ontology following from it only in chapter 13.

Machines and rules of rightness

Polanyi distinguishes among sciences referring to living and lifeless nature. He claims that these sciences have fundamentally different structures and explanatory systems.

> *"[...] only the physical sciences are predominantly observational, while biology and the study of mind and man have a more complex structure, in which observation plays but a subsidiary role."* (Polanyi 1958, 328)

Polanyi starts his argument by referring back to the evolutionary roots of humans that animals possess: 1) trick, 2) sign, and 3) latent learning (see ch. 5). According to him, there are in these three fundamental animal skills the roots of 1) contriving and engineering, 2) observing and natural sciences, and 3) understanding and mathematical and logical sciences. However, natural sciences include two main parts: on the one hand, physical sciences, referring to lifeless nature, and, on the other, life sciences, where the objective observation of the lifeless parts plays only a subsidiary part behind personal facts and judgments based on prior personal knowledge. (Polanyi 1958, 328) It follows that the recognition of an individual living being as a comprehensive whole is the exclusive act of personal knowledge. (Polanyi 1958, 343)

Before living beings, Polanyi examines machines in detail. According to him, a machine is nothing other than a complex, rational tool that we use during an act for the sake of some goal to reach some benefit. In this sense, its roots can be found in the animal knowledge of trick learning, which does the same only at a lower level and is the tacit basis of contriving and technology, i.e., engineering. The use and operation of a tool or machine are based on comprehensive rules of act and operational principles that cannot be determined by their details and that allow the successful usage of a tool or machine. It follows:

> "*The questions: 'Does the thing serve any purpose, and if so, what purpose, and how does it achieve it?' can be answered* only by testing the object practically as a possible instance of known, or conceivable, machines." (Polanyi 1958, 330)

On the contrary, the knowledge and recognition of an object or the material parts of a machine are based on the animal knowledge of sign learning, which is the tacit basis of observing and natural sciences. Therefore, these two forms of knowledge—practical engineering sciences and theoretical natural sciences—are highly different even in their tacit fundaments. Both are relevant regarding machines, but while engineering refers to the comprehensive whole and *operational principles* of machines, then physical sciences refers only to the tangible parts and *material conditions* of machines.

> *"A physical and chemical investigation cannot convey the understanding of a machine as expressed by its operational principles. In fact, it can say nothing at all about the way the machine works or ought to work."* (Polanyi 1958, 329)

> *"In other words, the class of things defined by a common operational principle cannot be even approximately specified in terms of physics and chemistry."* (Polanyi 1958, 329)

The machine is an achievement of an invention. The machine is defined by its patent that tries to describe the machine in the broadest possible sense; i.e., it does not include the concrete realization and the different possible material conditions of the machine, but only defines the operational principle—that is, the rules of rightness of the workings of the machine—due to which the machine can properly function and fulfill its goal. In this sense, a machine is an ideal with which several different concrete objects can comply. This ideal and these operational principles determine the concrete rules of rightness due to which the machine can work and accomplish its goals.

> *"The conceptions of machines in good working order form a system which ignores the particulars of failures—in the same way as geometrical crystallography ignores the imperfections of crystals. The operational principles of machines are therefore rules of rightness, which account only for the successful working of machines but leave their failures entirely unexplained."* (Polanyi 1958, 329)

However, the operation principles, according to their comprehensive nature, cannot answer why a machine starts malfunctioning, what is the cause of its failure—i.e., what is the concrete material part because of which the machine cannot function properly—in the same way as physical knowledge referring only to the material parts cannot tell in itself what a comprehensive machine is and why it is working. They are fundamentally different forms of knowledge, which complement each other.

Nonetheless, the relation of the two kinds of knowledge is not symmetric. On the one hand, if someone recognizes an object as a machine defined by a comprehensive function, then she has recognized the machine well—which cannot be claimed if she recognized it only as an object; i.e., she recognized only its material parts. On the other hand, however, physical knowledge is much more fundamental and can be applied to every object, while engineering knowledge can be applied only to those objects that are, at the same time, machines, too.

It follows that the two forms of knowledge are not only fundamentally different but can be linked together fruitfully. The detailed physical knowledge of a machine as an object can deepen the knowledge of the machine. Only engineering and its operational principles can account for the successful workings of a machine, but these operational principles and comprehensive knowledge, in themselves, cannot account for the failures of the machines because the cause of these failures can be found in the material conditions of the machines, which belong to the territory of physical sciences. Therefore, the failure of a machine can be accounted for only by these two different kinds of sciences together. However, learning the causes of the failures deepens the concept of the machine as a comprehensive whole, which finally can lead to new types of realization for the machine. Thus, although physical knowledge cannot account for a machine as a comprehensive whole, it is not unnecessary or optional at all but:

> *"[...] in our knowledge of a comprehensive entity, embodying a rule of rightness, any information supplied by physics and chemistry can play only a subsidiary part."* (Polanyi 1958, 331)

Successfully working machines are, therefore, the achievements of both engineering and physical sciences. Engineering determines the operational principles that account for the successful workings of machines, while physical sciences provide the material conditions among which the operational principles define the successful workings of machines. However, in case of failure, these operational principles of engineering only define a comprehensive framework in which, based on physical knowledge about the material parts, the cause of the failure can be determined. Thus, engineering accounts for the operational principles referring to the machine as a comprehensive whole, while physical sciences account for the proper material conditions and determine the possible cause of failures. (Polanyi 1958, 332)

Living beings and human reasoning

Contriving and observing, and thus engineering and physical sciences, about which Polanyi has talked up to this point, can be fundamentally characterized by the irreversible processes of human understanding. Now, Polanyi contrasts these with mathematical and logical sciences, which also have their roots in animal knowledge (latent learning) and can be characterized by reversible, deductive logical thinking.

> *"Thought proceeds largely by an irreversible process of comprehension and not according to specifiable rules. Only the latter will be called logical thinking, in which I shall include mathematics. Logic, thus defined, is a rule of rightness: it tells us how we must reason in order to derive correct and ample conclusions from given premises."* (Polanyi 1958, 333)

By this, Polanyi does not want to sharply contrast engineering and physical sciences with mathematical and logical sciences; he only wants to show the different nature of logical inference, and thus of logical machines, although they follow rules of rightness and concrete operational principles, as do machines. The rules of correct logical thinking have to be acquired in the same way as the rules of other forms of craftsmanship, and these rules cannot account for logical errors as the comprehensive operational principles of machines cannot account for the failures of machines. Psychology can account for the causes of logical errors. (Polanyi 1958, 333)

> *"Psychology cannot distinguish by itself between true and false inferences, and hence is blind to logical principles; but it can throw light on the conditions under which the understanding and operation of correct logico-mathematical reasoning may develop, and it may supply an explanation for errors in reasoning. Indeed, an error in reasoning can never be the subject of a logical demonstration; it can be understood only by psychological observations which reveal its causes."* (Polanyi 1958, 334)

This train of thought makes the same argument for the relation between logic and psychology, which we have seen in the case of the relationship between engineering and physical sciences. The conscious and unconscious processes of attention are the objects of psychology, while the principles of logic are based on these processes as "material" conditions. Polanyi refers to it but does not explain in detail that he also considers the principles of ethics and legal systems to be rules of rightness. (Polanyi 1958, 332)

It follows that there are different types of rules of rightness in all processes that are under the logic of achievement, i.e., that leads to achievements. In the case of machines, these are different types of operational principles; in the case of logic, they are different rules of inference—e.g., the modus ponens. In ethics, these are different rules of activities—e.g., "Thou shalt not kill." Although Polanyi does not mention it here, there are different rules of rightness in sciences, in arts, or in the case of the functioning of the human body (see ch. 12). The table below shows that which lower-level sciences in different areas account for the failures of higher-level principles.

Table 11.1: *Different sciences that complement each other*

Machines	Logic	Morals	Poetry	Living body
Physics, chemistry	Psychology	Sociology	Linguistics	Biophysics, biochemistry

The main difference between the relation between engineering and physical sciences and the relation between logic and psychology is that, contrary to machines, in the case of logic, there is a second person who tries to meet the rules of rightness, while a machine "only" works. The first person is the knower who, based on his own personal knowledge, can recognize that there is another person who is able to follow the rules of logic, i.e., who is capable of achievements. As we have seen, this comprehensive achievement cannot be recognized without the personal knowledge and judgment of the first person; i.e., based only on Laplacian, systematic, objective knowledge. (Polanyi 1958, 334)

The rules of rightness that a person tries to meet also set up an ideal in the same way in logic, legal systems, ethics, or in arts, as we have seen in the case of machines. This is the reason Polanyi says that a person has to commit himself to such ideals in every case, the medium of which is, in itself, blind towards these ideals. Physics and chemistry cannot account for the successful workings of a machine; that is, they are blind in terms of the rules of rightness according to which the machines work successfully. It follows that psychology is blind in terms of the rules of logic; sociology is blind in terms of the laws and the rules of act of ethics; linguistics is blind in terms of the maxima of poetry, etc.

At the same time, this blind medium and nothing else allows a person to act according to his ideals; in Polanyi's words: "... grants the possibility for striving for his ideal." (Polanyi 1958, 334) However, this blind medium also defines the conditions and constraints within which the person can act and gives as a lower-level foundation only a handful of clues for the person to meet his higher-level ideals. This contradictory situation determines a person's "calling" and leads to free, responsible actions. Contrary to this, a logical inference machine follows strict rules and has no responsibility for what it does; it only works successfully, or it fails. A human being, however, can be wrong at any time, and he has to bear the responsibility for his mistakes. (Polanyi 1958, 334)

> *"I accept the responsibility for drawing an ever indeterminate knowledge from unspecifiable clues, with an aim to universal validity; and this belief includes the acknowledgment of other persons as responsible centers of equally unspecifiable operations, aiming likewise at universal validity."* (Polanyi 1958, 336)

A person is not, therefore, merely a concrete tangible (material) being who can be explicitly specified by the sum of his parts through the methods of physical sciences or physiology. He is a comprehensive whole, a center based on his tangible (material) parts which have definite ideals, goals, and a calling that "dwell in" his different bodily and intellectual tools and his material conditions (see ch. 4. and BOX 9: Tacit integration). *The essence of a person is that he follows his ideal to which he committed himself.*

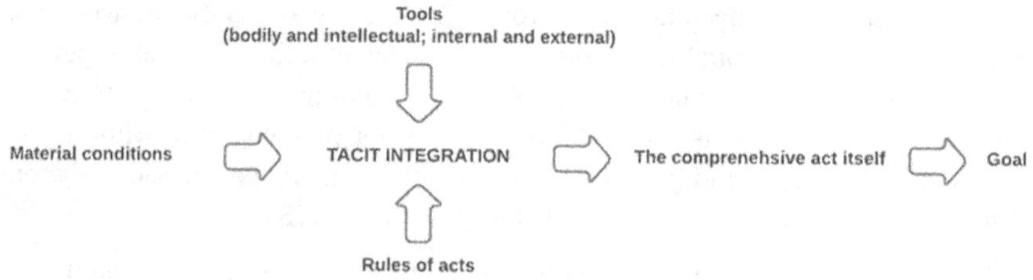

Figure 11.1: *Tacit integration*

Based on this, Polanyi asserts:

> *"To represent living men as insentient is empirically false, but to regard them as thoughtful automata is logical nonsense. For we are aware of a man's thoughts only by listening to him, i.e. by attending subsidiarily to certain bodily actions in the assumption that they are impelled by his thoughts, which are in fact known to us only as the effective center of his meaningful actions. Nor can we speak therefore of thinking which totally lacks originality and responsibility."* (Polanyi 1958, 339)

Unfortunately, neurology focally observes only the lower-level bodily processes and regards human thinking as the analogy of a single-level logical inference machine. It follows that it ignores the personality, the responsibility, and the calling of human beings, and it also ignores the active, creative power of human beings that they inherited from their earlier animal life. (Polanyi 1958, 339) We can find the roots of the human person and personal knowledge not in its material parts or a logical inference machine but in animals, animal knowledge, and in the evolutionary development of human beings. This is the reason Polanyi states that inside us, there is "an active center operating unspecifiably in all animals." (Polanyi 1958, 336) The body of an animal, in itself, is constructed from machine-like tools, the successful workings of which are determined by fixed rules of rightness, as we have seen in the case of machines. This, however, would not be enough for animals in real evolutionary challenges. Machines also cannot work if humans do not build and operate them. However, contrary to machines, in animals, there are tacit, creative powers that activate these fixed operational principles and machine-like bodily tools. We can find the essence of animals in the center of these tacit, creative powers. This governs the body of the animal constructed from *machine-like tools.*

> "[...] *while the animal's machinery embodies fixed operational principles, this machinery would be impelled, guided and readapted by the animal's unspecifiable inventive urge* [...]" (Polanyi 1958, 337)

These are the tacit, creative powers that are also operating in the case of human beings. Humans, however, have not only the bodily tools that animals have, but, with their intellectual and external tools, they can significantly unfold the tacit, creative powers they inherited from their animal life. Just think back to the trick, sign, and latent learning by animals, which, through man's intellectual and external tools, become invention and engineering, systematic observation and natural sciences, inference, and mathematical and logical sciences.

> *"The evolutionary process forms a continuous transition from the inanimate stage to that of living and knowing persons."* (Polanyi 1958, 345)

This leads to the exact same relation between biology and physical sciences as we have seen between logic and psychology and between engineering and physical sciences.

> *"Morphogenesis is the formation of right shapes, a process which may succeed or fail. A physical-chemical explanation would not account for these alternatives, it would merely shift the problem back to the rightness of the conditions from which the process started.* [...]
> *Studies of physical-chemical processes can never take the place of these interests; they can belong to psychology or embryology only to the extent to which they have a bearing on anterior interests arising within these sciences. Physical and chemical knowledge can form part of biology only in its bearing on previously established biological shapes and functions: a complete physical and chemical topography of a frog would tell us nothing about it as a frog, unless we knew it previously as a frog."* (Polanyi 1958, 342)

The active, creative powers of animals (see ch. 6) therefore initiate and control the different and fixed machine-like tools in an animal's body. This control can be specified only through comprehensive concepts. The precise operational principles of machine-like tools can be understood through destructive analysis (see ch. 4), but the appraisal of these comprehensive operational principles can be conducted only on the basis of personal skills. (Polanyi 1958, 342) The key step is the recognition of the animal's *center* and *individuality*, which differentiate between machines and animals. According to Polanyi, this is the exclusive act of personal knowledge since it is about personal and not objective facts. (Polanyi 1958, 344)

It follows that we have three logical levels in the case of biology and life sciences. There is our own personal perspective and knowledge; there are the lifeless material parts or conditions of physical sciences, and the most important part is the recognition and acknowledgment of the center and individuality of living beings. (Polanyi 1958, 344) Contrary to the objects of physical sciences, this last factor fundamentally transforms our relationship to living beings and biology. With regard to the examination of human beings, our "I/It" relationship to the objects of physical sciences entirely becomes an "I-Thou" relationship. According to Polanyi:

> *"This suggests the possibility of a continuous transition from statements of fact to affirmations of moral and civic commands."* (Polanyi 1958, 346)

Conclusion

There are comprehensive entities, certain acts of which can be understood only in relation to their success or failure. This means that the given act is evaluated based on its rules of rightness, which determine its success. If the given act complies with its rules of rightness, then it has to be acknowledged as the achievement of the comprehensive entity. Without the rules of rightness, it cannot be decided that an act is a success or a failure because, from the perspective of the tangible parts, all acts are the consequence of the mechanical processes of these parts.

Since physical and chemical examinations involving only the material conditions of comprehensive entities do not include any rules of rightness, these kinds of examinations are not capable of appraising the achievements of comprehensive entities. Thus, if we ultimately insist that only physical and chemical types of examinations are acceptable in science, then we will question the real achievement of comprehensive entities and living beings. This would also lead to our questioning the reality of living beings since living beings can be differentiated from simple material systems only based on the (personal) fact that they possess knowledge through which they can realize real achievements.

The recognition and appraisal of successful efforts of living beings is also an achievement of this type, which is based on the tacit skills and personal judgments of human beings. Biology (or engineering)—the assignment of which is the recognition (or discovery) of different rules of rightness and the appraisal of the achievements of living beings—is a fundamentally different science from physics and chemistry, which are directed only towards the material parts and conditions of any object and living being. Unfortunately, this fact was not acknowledged by critical philosophy.

Chapter 12

Knowing life

Goals of the chapter

1. Polanyi connects the results of his investigation on the nature of tacit knowledge with his claims on biology and calls this step the "point of confluence" where the two different branches of his argument unite. (Polanyi 1958, 379)
2. He asserts that biology is but life reflecting on itself. Since the biologist is part of living nature, he (with his tacit skills and knowledge) is also the object of biology. (Polanyi 1958, 374)
3. This recognition urged him to undertake "a systematic survey of the tacit coefficient of knowing, as it ascends and broadens out from the primordial activity of the lowest animals to the whole edifice of human thought in human society," actually, to write *Personal Knowledge.* (Polanyi 1958, 373)
4. One of Polanyi's main goals is the critique of contemporary biology. He points out the inadequacy of several contemporary approaches, especially of behaviorism, because in these cases, the starting hypotheses and the explanatory principles cannot be reconciled with the tacit fundaments of knowledge.
5. The adequate biological approach is not based on the same hypotheses and principles as physics and chemistry, and it also cannot be cultivated by the methods of physics. Similar to engineering, biology has to deal not with the objective facts of the material parts but with the personal facts of comprehensive wholes (see ch. 4). These facts can only be described by the normative concepts of the logic of achievement and not by the neutral terms of physics and chemistry (see ch. 11).
6. Since biology is a kind of knowledge the object of which is knowledge itself—ascending from the lowest biotic levels to the highest intellectual activities of human science—we have to speak about an expanded concept of biology, referred to by Polanyi as "ultrabiology," and not only about mechanical biology based on the concepts of neo-Darwinism and behaviorism. (Polanyi 1958, 363)

7. While the objects of physical sciences—e.g., stars, crystals, electrons, etc.—sharply separate both from physical sciences and personal knowledge, the objects of biology, e.g., ecosystems, apes, eukaryotes, etc., with their primitive functions, achievements, and commitments, are all integral parts of higher-level human personal knowledge.

***

"This is biology. It is a participation of the biologist in various levels of commitment of other organisms, usually lower than himself. At these levels he acknowledges trueness to type, equipotentiality, operational principles, drives, perception and animal intelligence, according to standards accepted by him for the organisms in question." (Polanyi 1958, 379)

"Facts about living things are more highly personal than the facts of the inanimate world. Moreover, as we ascend to higher manifestations of life, we have to exercise evermore personal faculties—involving a more far-reaching participation of the knower—in order to understand life." (Polanyi 1958, 347)

Taxonomy

First of all, Polanyi shows that there are true living forms, or types, the recognition and classification of which are based on our personal knowledge, on our basic tacit skills. Polanyi, of course, speaks of forms, not in the Aristotelian sense or based on the genetic definition of neo-Darwinism, but he defines them by their operational principles and achievements.

The appraisal and classification of living types called *taxonomy* is an ancient, descriptive science, the basic skills of which everybody practices in his everyday life. We recognize a cat as a cat, a frog as a frog, etc. Moreover, these basic skills are possessed by the animals themselves: a cat can recognize a dog and can recognize a mouse, and the result determines its further actions. The fact that we recognize a cat as a cat means that we acknowledge it as a real being. (Polanyi 1958, 348)

A given living form cannot be exactly defined. (Polanyi 1958, 348) Those who recognize it do not do so due to an exact definition but due to a tacit integration that leads to the recognition of the given comprehensive living form from the subsidiary awareness of its characteristic parts (see ch. 4 and 11). Also, when we recognize it, we commit ourselves to the belief that there is a living form—a given individual whom we have just recognized. We appraise and categorize the different living forms into classes based on rational criteria, and we expect these classes to be true. (Polanyi 1958, 349)

The criteria of a biologist are different from the criteria of, e.g., crystallography, in the sense that the former is based on *experience*, i.e., on our previous tacit impressions of living beings, while the latter is based on *ideals*; i.e., on geometric forms. Every classification of a concrete living being into a living form includes the small *reappraisal* of that class. Its degree, of course, changes significantly in the case of a new species compared to a well-known one. But this happens in every case because the classes were constructed and are continuously reconstructed on an empirical basis vis-à-vis these concrete individuals, while the classification of a crystal into a given crystal class does not change the given class at all. (Polanyi 1958, 349)

The classification of plants by Carl Linnaeus was an artificial system based on ideals, yet Linnaeus himself already tried to construct a natural system based only on experience about living beings. His goals were finally reached by Charles Darwin, who showed through his *On the Origin of Species* that the classification of living forms has a deeper meaning: there are no definite, distinct classes and sharp boundaries in the living world which could comply with some external ideal criteria, but living classes gradually evolve from each other. (Polanyi 1958, 349-350)

Taxonomy has had great results ever since, but it is still not greatly appreciated in science. The reason for this, according to Polanyi, is that taxonomy is based on *connoisseurship* and has a descriptive nature. Since living forms cannot be exactly defined as crystal classes, the recognition and categorization of living beings, even in the highly explicit scientific framework of taxonomy, strongly relies on those deeper, everyday tacit skills by which someone can recognize a cat as a cat or a frog as a frog. This kind of competence and use of tacit skills is not present in physics or population genetics.

> *"One might expect to find this grandiose achievement celebrated wherever biology—the science of animals and plants—is taught and cherished. But no; classical taxonomy has almost ceased to count as a science. The explanation seems to lie in a change in the valuation of knowledge. It is due to a steadily mounting distaste for certain forms of knowing and being; a growing reluctance to credit ourselves with the capacity for personal knowing, and a corresponding unwillingness to recognize the reality of the unspecifiable entities established by such knowing."* (Polanyi 1958, 350)

Since taxonomy and the comprehensive living forms described by taxonomy do not comply with the objectivist ideal of knowledge, science starts to doubt both the reality of living forms and the fact that taxonomy is real science like physics or population genetics.

Nonetheless, this aversion does not result in the consequent denial of the existence of living forms in practice—this is stated only at a theoretical level—but merely tries to describe living forms by the less personal concepts of genetics. Biology would be completely destroyed if we started to regard living forms as nonexistent in practice, too, because the genetic examination of a given population, in fact, also *presupposes* that this population can be separated from other populations based on taxonomy. Otherwise, we would not be able to give the exact genetic description of that population—at most, only the description of *all* populations—since genetics in itself has no reason to classify the different living beings, which are all connected by continuous genetic transitions, into different taxonomical categories based on morphological differences.

> *"To undertake genetic experiments irrespective of morphological differences, with the intention of establishing from these alone the boundaries of specificity, would be absurd. It has certainly never been contemplated."* (Polanyi 1958, 353)

> *"Projects for the application of additional, and in particular more objective, taxonomic tests, have all set themselves their tasks within the existing morphological system."* (Polanyi 1958, 353)

Polanyi, of course, does not want to claim that the detailed investigation of the more objective parts—that is, the critical destructive analysis of a comprehensive whole; e.g., a living form—could not lead to the result that the comprehensive whole in question is only illusory, like the objects of astronomy or Cassiopeia's W-scheme (see ch. 3). A successful destructive analysis can even destroy our genuine interest in an object which has been identified as something illusory. For instance, in the case of living forms, suppose that genetics without the concepts of taxonomy was able to speak about those comprehensive forms that had been recognized falsely as individuals of certain species by taxonomy. This result would be similar to the case of the illusory objects in astrology when astronomy became able to speak consequently about the celestial phenomena—e.g., the movements of planets—without the false concepts of astrology. If genetics, in the case of living forms, cannot reach a similar consequent result, then genetics cannot be regarded as the complete explanation of living forms. (Polanyi 1958, 389)

> *"If consistently carried out, the denial of contemplative value in science would cut off biology from the intellectual passions from which it takes its origin, and could not stop short of denying altogether scientific reality to the beings in which life manifests itself."* (Polanyi 1958, 354)

According to Polanyi, in reality, no one wants to reach this result, since tacitly, we unquestionably acknowledge that living beings are the individuals of their species. In practice, genetics is also based on the personal facts and concepts of taxonomy without any problem. No one believes seriously that our passionate interest in living beings will evaporate someday. Nonetheless, we explicitly state our doubt again and again both in personal knowledge and the reality of living forms, and, for the sake of a false scientific ideal, by deceptive substitutions, we conceal what we indeed tacitly believe (see ch. 1). According to Polanyi, this behavior is a kind of modern intellectual equivocation by which we are flattering our "sense of intellectual superiority." (Polanyi 1958, 389)

Morphogenesis

Polanyi shows that the formation of living forms, or types, is a biotic achievement that cannot be described by the neutral concepts of physical sciences.

Polanyi sheds light on the tight connection of reproduction and regeneration underlying the graduality of evolution. Regeneration is the recovery of the mutilated organization. In the case of a primitive organism, it is *equipotent*, which means that the whole organism is able to regenerate. The replication of a unicellular organism can be regarded as a kind of (self-) mutilation that is followed by the process of regeneration. More exactly, mutilation and regeneration are the two composite parts of the process of replication, which, in turn, is the most fundamental case of reproduction. (Polanyi 1958, 355) However, in the case of more developed organisms, this kind of full equipotency is replaced by the system of *fixed potency* of machine-like tools, which means that the processes of replication, reproduction, and regeneration are finally separated at different organizational levels. Thus, regeneration works only in certain fixed, local subsystems. If we cut our finger, it regenerates, but if we cut off the whole finger, it does not.

Morphogenesis is the comprehensive process, as the different fixed machine-like tools of the organism develop during reproduction, which in a normal case happens in a coordinated manner in the whole organism due to the controlling mechanisms of the organism's center. This comprehensive achievement during the morphogenesis of the whole body cannot be defined by the fixed potency of the subsystems. For example, there are no genetic differences among the genomes of the cells in different fixed subsystems, but there are general morphogenetic (operational) principles that initiate the workings of different dormant genes in those different fixed subsystems. (Polanyi 1958, 357)

Taxonomy, as we have seen previously, defines the different living forms based on basic tacit skills; morphology describes the specific structures of living beings, while embryology studies the basic formation of these structures and defines the normal stages of the development of living beings. Experimental embryology, concerning primarily the conditions of this development, is usually regarded as a successful and high-value science. Polanyi, however, emphasizes that there is a hierarchy among experimental embryology, descriptive embryology, morphology, and taxonomy since they are based on each other. Therefore, through descriptive embryology, morphology, and taxonomy, even experimental embryology is based on our basic tacit skills and personal facts. Nonetheless, the main question is in which conditions and due to which comprehensive principles can the embryos meet the criteria of these normal stages of morphogenetic development defined by morphology. (Polanyi 1958, 356) According to Polanyi, the role of morphogenetic principles is "to define operations for the achievement of ontogenetic success conceived in morphological terms." (Polanyi 1958, 357) This means that the development of living beings can be appraised due to *success* or *failure*; i.e., the living individual will be normal or degenerate. Therefore, the development of living beings complies with the *logic of achievement* and not with the neutral terms of lower-level mechanical processes.

> *"The meaning of experimental embryology is thus doubly dependent on personal knowledge: both in respect of the unspecifiable knowledge of true shapes, and in respect of the appreciation of the process by which highly significant shapes and structures are brought into existence."* (Polanyi 1958, 357)

We have seen that taxonomy and morphology heavily rely on personal facts. This reliance is deepened by the personal appraisal of the dynamic processes that form the specific structures of living beings as they reproduce. These processes will only realize this basic goal in the case of living beings who comply with each normative criterion of the stages of development. This fact causes a level of uneasiness among scientists, in the same way as we have seen in the case of taxonomy. (Polanyi 1958, 358)

Although experimental embryology has highly precise measurement and experimental methods to collect exact data and uses advanced mathematics to deal with that data, it cannot be more objective because of the normative framework in which each developmental process takes place.

> *"Normal shapes—as distinct from abnormal, malformed, stunted shapes—would have to be identified by our own standards of rightness before they could be defined in mathematical terms. Mathematical relations are, like the processes of physics and chemistry, neutral in respect to morphogenetic success or failure; these alternatives must therefore be identified by ourselves before we can analyze them in terms of mathematics or physics and chemistry."* (Polanyi 1958, 358)

Since mathematical formulas are neutral regarding success or failure, any success or failure of living beings has to be appraised prior to the application of the exact formulas, based on our personal knowledge.

Polanyi once again mentions his argument concerning Laplace's perfect intellect (see in detail in ch. 6) that here we are dealing with comprehensive morphological and morphogenetic schemes that cannot be appraised or even recognized by objective, impersonal knowing.

> *"Indeed, a complete 'Democritean' or Laplacian knowledge can tell us nothing without relying on our personal knowledge of these comprehensive features. Suppose we were given a complete topographical chart of all physico-chemical changes taking place in our surroundings. It would require a superhuman feat of insight to discover from this information the fact that there are somewhere such things as chickens, which are hatched from eggs."* (Polanyi 1958, 358)

Laplace's demon, without tacit bodily skills and personal knowledge, would not be able to recognize these comprehensive facts like chickens and eggs.

> *"We may conclude that the insights by which we recognize life in individual plants and animals, and distinguish their several kinds—and by which we appraise them as normal or abnormal, establishing thereby the success or failure of the process by which they come into existence—that these insights reveal a reality to which we have access by no other channels, and that the mechanism of morphogenesis can therefore never amount to anything but the observation and understanding of patterns and processes expressly bearing on that reality."* (Polanyi 1958, 359)

In a sense, living beings are machines; however, they are living machines—that is, although they are in the same comprehensive class as artificial machines, they are, neither in their appearance and behavior, nor in their essence, identical with machines. Therefore, one of the main subclasses of machines, *living machines,* includes living beings, while the other, *artificial machines,* includes well-known man-made instruments. In the case of artificial machines, we can speak of further lower-level subclasses:

1. Simple machines, like a hammer;
2. Normal machines, like a mechanical clock;
3. Complex machines that have centers, like an autonomic humanoid robot[1].

However, living beings are all complex and independent, and, most of all, they include a center (see ch. 4).

BOX 21: The theory of boundary conditions

Polanyi details his claim that living beings belong to the same comprehensive class as machines in his *Science* paper, *Life's Irreducible Structure.* They belong to the same class because both the structure of living beings and the structure of machines can only be described by a hierarchy of comprehensive orders ("boundary conditions"), which comply with certain operational principles and the proper functioning of which can only be understood by the normative terms of the logic of achievement. Contrary to these structures, the comprehensive structures of lifeless natural objects such as crystals or rocks do not comply with any operational principle since these structures are merely the consequences of lower-level material/mechanical processes. Machines were created by man based on the principles of engineering (see ch. 11), and living beings emerge due to the principles of life and evolution (see ch. 13). Based on this fact, Polanyi differentiates between *machine* (e.g., living beings and artificial machines) and *test-tube* (e.g., crystals) type boundary conditions. So although machines and living beings have a different ontological nature, both comply with operational principles, and both function due to the logic of achievement; that is, they belong to one main subclass of machine type boundary conditions.

---

[1] For a detailed analysis on this topic, see Héder, Mihály and Dániel Paksi. "Autonomous Robots and Tacit Knowledge". *Appraisal: The Journal of the British Personalist Forum* 9: 2, 8-14, 2012.

Since living beings are living machines, the relation of physiology to physics and chemistry is the same as that of engineering to these fundamental sciences. "Physiology is a system of rules of rightness, and as such can account only for health." (Polanyi 1958, 360) The main difference between the logic of physiology and the logic of engineering is that living beings usually can realize their goals and functions in multiple ways, while in machines, these goals and functions are fixed. This fact is the reason Polanyi states that the unspecifiability of physiology is twofold: neither the operational principles and rules of rightness of the functions, nor their realizations, can be exactly determined. However, the point is the same also in this case:

> *"Any attempt to conduct physico-chemical investigations of a living being irrespective of physiological assumptions will lead as a rule to meaningless results; and any attempt to replace physiology altogether by a physico-chemical chart of the living organism would completely dissolve any understanding of the organism."* (Polanyi 1958, 360)

Action and perception

Polanyi shows that animals have active centers and commitments in terms of reality, and they act according to their commitments, which also can only be appraised by the normative terms of the logic of achievement.

Polanyi distinguishes between the concepts of *action* and *perception*. This is important because if we do not do that, then the free actions of living beings easily become merely determined reactions of perceived external stimuli. He uses the concept of perception in the usual sense; that is, perception is the process by which an external object evokes certain impressions in the sense of a living being, evoking the emergence of certain knowledge that has denotative meaning concerning the given external object. Perception is "consciously discriminating"—in the sense of subsidiary awareness—but not able to actively discriminate yet. (Polanyi 1958, 361) This means that not all external effects on the senses evoke the emergence of new knowledge; however, in terms of which effects are ignored and which become the sources of new knowledge, this is not the result of a deliberate decision. According to this, perception follows self-defined norms and creates general conceptual schemes in which the given living being understands its environment; e.g., objects maintain their sizes and shapes. (Polanyi 1958, 96-97)

> *"This tremendous generalization, on which we base our understanding of the universe, we have in common with the higher animals. Their native sensory equipment and ours both set us similar standards of correct seeing."* (Polanyi 1958, 97)

Therefore, perception, the most important tacit basis of, for example, taxonomy, already includes tacit concepts related to reality.

Polanyi regards action as something active and deliberate that is the consequence of conscious motivation. According to this, action, contrary to perception, is not just the function of an organ but the *act of an active center.* (Polanyi 1958, 363)[2]

BOX 22: Polanyi's use of consciousness

Polanyi, of course, does not use the term "conscious" in its original, narrow meaning as determined by the concept of dualism, according to which solely humans can be conscious, but in its broader, emergentist sense, according to which any living being is conscious to a degree. For example, when, as a result of successful recognition, a cat runs away from a dog, it is perfectly aware of the presence of the dog (a conceptual category including certain individual living beings), and it deliberately and consciously chooses to flee rather than fight. Consciousness is not a separate quality of man but a gradually developing tacit skill in evolutionary emergence (see ch. 13). Nonetheless, it is true that, due to our new information recording and transmitting skills, we have far more developed consciousness than even primates.

Both perception and action can lead to failure.

1. At the level of existence (the category of which is defined by taxonomy), failure is an *abnormality.*
2. At the level of development (the stages of which are defined by morphology), failure is *deformity.*
3. At the level of functioning (like perception), it is *disease* (e.g., if someone is color-blind).
4. At the level of the active, perceptive person, there are two kinds of failure:

[2] For a more elaborate description see Héder, Mihály and Dániel Paksi. "Non-Human Knowledge According to Michael Polanyi". *Tradition and Discovery* 44: 1, 50-65, 2018.

a. *Subjectivity;*

b. *Error.*

Since we are speaking of success or failure, both perception and action follow the logic of achievement, which can only be appraised by personal participation and knowledge, as we have seen in the cases of machines, taxonomy, and morphogenesis. That is, using only the methods and concepts of physics and chemistry, we can only see the random motions of material parts and cannot account for success and failure, which have a comprehensive nature. However, the causes of failure here can also be found at the lower material/mechanical level, as we have seen in the case of machines (see ch. 11). (Polanyi 1958, 361)

Polanyi's most detailed example is nutrition: when a fish bites on bait, the result is not what it expected, what it intended to do. We can judge what happened as a failure only if, on the one hand, we define what the *normal nutrition* is for the fish and, on the other hand, we suppose that there is a *rational, active center* in the fish, and what happened was the *deliberate act* of this rational, active center. (Polanyi 1958, 362) This can be done only by our personal participation because we have to decide, in the context of the comprehensive system, what is the bait and what is normal nutrition for the fish. Based only on the concepts and methods of physical sciences, we can only see that the mechanical processes of the material parts as a complex material system (i.e., the fish) react to certain stimuli. However, the causes of the failure can be found at the lower level; that is, in the material conditions, and not in the higher-level principles according to which the fish tries to act successfully.

If we acknowledge that this was the deliberate act of a rational, active center, then it has another consequence. In Polanyi's words:

> *"To swallow something in the hope that it may be wholesome is clearly a commitment, and so is every act of seeing things in one particular way."* (Polanyi 1958, 363)

The fish divides things into two: nutrition and non-nutrition. However, for the fish, it is not just a theory since it cannot construct explicit theories but rather *a tacit commitment* towards reality, according to which it acts. Therefore, the fish acts according to its tacit categories and commitments, and we judge this act according to our normative categories: when it bites the bait, we state that it made an error.

> *"I have suggested before that in a generalized sense commitment may be acknowledged even at the vegetative level, since it is of the essence of a living organism that each part relies for its function, and for its very meaning as part of the organism, on the presence and proper functioning of a number of other parts. In this sense our knowledge of the normal growth, functioning and being of the organism is an appraisal of its primordial commitments which accredits their success."* (Polanyi 1958, 363)

So the appraisal of the achievements of living beings is also the appraisal of their commitments towards reality. Then, when we start to appraise active living beings who can do biology, we have to state that biology is "a responsible commitment that appraises other commitments." (Polanyi 1958, 363)

The degrees of commitment by which consciousness increases, according to Polanyi, are the following: (Polanyi 1958, 363)

1. The *vegetative* commitments of a center of *existence, growth,* and *function;*
2. The *primitive* commitments of a *perceptive* center;
3. The *responsible* commitments of the *consciously deliberating person.*

According to this evolutionary gradualism, Polanyi does not restrict biology only to the appraisal of lower-level commitments but expands it into "ultrabiology" that also takes on the appraisal of the higher-level, responsible commitments of human persons. (Polanyi 1958, 363)

The active-perceptive center possesses the following main tacit skills: awareness of motivation and knowledge; effort to do right and know truly; a belief in an independent reality; and a sense of hazards of acts. (Polanyi 1958, 363)

The classes of appraisal (normality vs. abnormality or deformity, and health vs. decease) are extended due to sentience in the following way: (Polanyi 1958, 363)

1. Correct satisfaction of normal criteria: e.g., the fish bites on food;
2. Incorrect satisfaction of normal criteria—that is, error: e.g., the fish bites on the bait;

3. Action or perception satisfying subjective, illusory criteria: e.g., the fish has impaired sight, so it recognizes certain non-nutrition as food and bites on it;
4. Mental derangement issuing in meaningless reactions: e.g., although it can see that it is not nutrition, the fish obsessively eats sand.

Polanyi formulates against behaviorism that we do not have to beware of dwelling in, for situations of living beings, in order to preserve an objective standpoint. On the contrary, we can better understand their actions, motivations, beliefs, and personality in this way. (Polanyi 1958, 364)

Learning and knowledge

Polanyi argues that learning and problem-solving follow the logic of action and perception. Since both learning and problem-solving are personal achievements, they can neither be described by the impersonal, objective lower-level physical-chemical concepts and principles, nor by a higher-level, but also impersonal, so-called objective theory of knowledge.

Classifying empirical inferences, Polanyi finds the same four categories as above in the cases of action and perception. (Polanyi 1958, 366)

1. *Rational inference*—correct inference in a true system;
2. *Rational error*—incorrect inference in a true system;
3. *Subjectively rational inference*—correct inference in a false system;
4. *Incoherence*—there is no system, or rational inference.

Naturally, (case 3b) there are subjectively rational errors in false conceptual and explanatory systems, too, but the result of the process is, of course, exactly the same as in our original case 3; that is, error.

> *"For a problem is the intimation of a hidden rational relationship which is felt to be accessible by an heuristic effort, and the discovery of which may be accompanied, even in animals, by the lively enjoyment of their own ingenuity. By searching for such a hidden relation and by its eventual joyful acceptance, the animal reaches out towards something more objectively satisfying than food or sex, and in this sense the ensuing commitment becomes more radical."* (Polanyi 1958, 366-367)

Polanyi states that there are three parallel shifts in emphasis as we proceed from case 4 towards case 1; that is, from incoherence towards the unfolding and understanding of a hidden and real rational interrelation. Accordingly:

1. The *need for universality* is strengthening.
2. The *heuristic impulse* is also strengthening.
3. As a result of the first two, the *act of commitment* is strengthening, too.

On the one hand, these processes describe the gradual deepening of personality. On the other, there are existing comprehensive and rational relations (patterns, operational principles, interactions, etc.) in reality, which can be recognized and understood by the tacit skills of animals. These heuristic achievements give the animal more satisfaction than satiating its appetitive or sexual desires because, through the solutions to these problems, it gains real, lasting knowledge that changes its life—contrarily, appetitive or sexual desires always return in the same form. Polanyi's point is that, by problem-solving, the animal can commit itself to a reality that is more comprehensive, more universal, and has deeper existential meaning than random fundamental material processes have. According to Polanyi, this is a personal act that can only be appraised by the more personal participation of the biologist. (Polanyi 1958, 366-367)

It follows that psychology, like morphogenesis or action and perception, also follows the logic of achievement; however, in contrast to those, it is more profound and personal.

> *"And this is why psychology differs from astronomy; it does not describe events related to no purpose, but analyzes a certain class of achievements believed to be mental. The result is a system of rightness, which depends on certain not normative elements for its success or failure."* (Polanyi 1958, 369)

However, psychology, due to the mechanical vs. vitalism dichotomy—which is one of the applications of the more general objective vs. subjective dichotomy—doubts the reality of any kind of teleology. Consequently, psychology regards any teleology as vitalist obscurantism, preventing the consequent use of the scientific method. Polanyi, due to the logic of achievement, harshly criticizes this approach, since psychology, in this way, regards animals as mechanical inference machines in which there are no skills for real heuristic achievements, so they try to understand the results of learning and problem-solving based on mechanical principles and random mechanisms, such as trial and error, which can be modeled by logical rules. Consequently, this approach presupposes that learning theories can be separated from psychological processes. (Polanyi 1958, 371)

Nonetheless, this separation can only be made in the case of mental processes such as mathematical-logical inferences, which have well-defined operational principles and a very high degree of explication and formalization, and for which personal participation in the process can be reduced to a minimum degree. However, these steps cannot be taken in the case of empirical-inductive inferences because these inferences follow loosely-defined, vague operational principles and maxims that can only be effectively and successfully applied through exceptional and well-practiced bodily and intellectual skills. Therefore, in this case, the vague object of learning theory cannot be well-separated from the simultaneous investigation of lower-level psychological processes; that is, learning theory should remain part of psychology and thus indirectly of biology.

Since in a separated learning theory, the modeling of the lower-level psychological processes is, of course, done mechanically, at the higher level, these—at the lower-level deterministic—psychological processes are manifested as random events (see ch. 3) that are treated by different formal solutions (such as trial and error) by the higher-level logical inference-system. However, this handling does not comply with the logic of achievement since it entirely ignores the teleological nature of learning and problem-solving.

> *"For even the most elaborate objectivist nomenclature cannot conceal the teleological character of learning and the normative intention of its study. Its supposedly objective terms still do not refer to purposeless facts but to well functioning things."* (Polanyi 1958, 371-372)

Consequently, in these kinds of separated learning theories, for the sake of an objectivist ideal, the role of the animal is not acknowledged in its own achievement.

Superior knowledge

With regard to superior (human) knowledge, Polanyi emphasizes that, without the acknowledgment of the active personal center that can make heuristic achievements and commitments, we cannot even account for our own beliefs, knowledge, and commitments.

The territory of superior knowledge can be found where ultrabiology, that is, the investigation of cultural phenomena and their tacit roots—starts to work. This is the territory where the biologist, in appraising the different kinds of commitments towards reality, meets his own responsible personal commitments in the case of other human persons. Here, the *I-It* (person-object) relationship of any scientific inquiry is replaced by the more personal *I-You* (person-person) one. This change presupposes such communicative acts in the framework of which the participants mutually question, inform, judge, and persuade each other. (Polanyi 1958, 375)

These communicative acts among equals lead to the establishment of different social contracts (consensuses), which mean the social and moral commitment of the community towards a new aspect of reality. However, it is also possible at this cultural level to make misstatements or to infer falsely due to the previously detailed categories. In Polanyi's view, these higher-level social commitments and consensuses are not more fundamental than the lower-level commitments of biology. Social commitments, creating a more personal level, are always built on the former, lower-level commitments, but this level also entirely follows the logic of achievement. (Polanyi 1958, 375) The four levels are as follows:

1. *Vegetative;*
2. *Primitive;*
3. *Personal;*
4. *Social.*

Superior social and cultural knowledge is not preserved in its entirety since that would be impossible. Most scientific papers, novels, legal works, etc., are only gathering dust in bottom drawers; however, we can find the most important conclusions of these works in certain distinguished books, in certain highly esteemed persons' teachings, and we follow these conclusions and teachings.

> *"When referring to such superior knowledge we are not laying down standards for judging the persons to whom we attribute this knowledge; we are submitting, on the contrary, to the standards laid down by them for our own guidance."* (Polanyi 1958, 376)

These distinguished teachings and knowledge lead us to the fifth major level of commitments, which concerns those shared fundaments on which the communication and the establishment of consensuses are grounded at the social and cultural level. "A responsible encounter presupposes a common firmament of superior knowledge." (Polanyi 1958, 378)

This hierarchical structure is similar to those relations that we have seen in the case of craftsmanship learning (see ch. 4). The learner commits himself to the authority of his masters and uncritically accepts and follows their lead; that is, the process of learning is essentially tacit. However, acquisition of superior knowledge takes place in a highly articulated conceptual system and is led by the explicit communicative acts of the learning community where the acknowledged authority is not manifested in any concrete persons, but in the ideals, truth, and reality they present. (Polanyi 1958, 375-376)

> *"For it is not to their person, but to what we understand to be their teaching, that we pledge ourselves. It is indeed only by the lives of ordinary men within a free society that the principles to which it is dedicated acquire their effective meaning. The superior knowledge guiding a free society is formulated by its great men and embodied in its tradition."* (Polanyi 1958, 377)

Polanyi looks back to ch. 7 to argue why he speaks only about free societies that, on the one hand, acknowledge thinking as an independent force—otherwise, the communication acts of the fourth level would be highly different—and, on the other hand, at the fifth, highest level of commitments, are still governed by authority and tradition. (Polanyi 1958, 376)

In Polanyi's words, this is the point of confluence where the expanded ultrabiology meets the commitments of a free society to the intellectual norms of its culture, where the need for universality embedded in our centers reaches its goal in the gradual unfolding and understanding of reality by free thinking. Here is the point where the two branches of his argument, the epistemological and the biological, are inseparably interwoven into each other.

As we have seen, there is an intention during problem-solving already in the case of animals to reach higher-level knowledge through which they grasp universal aspects of reality that are beyond their primitive personhood and reality. However, at this level, this intention and need for universality cannot yet be articulated and shared, which hinders the emergence of higher-level commitments that go beyond the reality and personhood of the animal but which, in an articulated culture, would lead to the common acknowledgment and following of general ideals and truths. Therefore, this can happen merely in the case of man based on his articulated culture and communication—though now the person cannot be found at the higher pole of his commitment, but, on the contrary, he *subordinates himself* to impersonal, universal truths and principles.

> *"He may be said to stand on the lower level of this commitment. Alternatively, we may describe him as forming the personal pole of a*

commitment of which the ideals of man form the universal pole." (Polanyi 1958, 379)

Conclusion

In this chapter, Polanyi heavily builds on the logic of achievement we have just seen in the previous chapter. He sharply separates biology from physics and chemistry and, according to the notion of evolution, expands it in the territory of higher-level human knowledge, including the reflective investigation of biology itself (ultrabiology).

With regard to the logic of achievement, biology significantly differs from engineering because it investigates beings that have active centers, and it includes several hierarchical levels built on each other. At all of these levels, new achievements, new knowledge, and new commitments arise, all of which have their own operational principles and rules of rightness. The biologist can account for these evermore-personal achievements, successes, and failures only by his personal participation and appraisal of these commitments. The objectivist conceptual scheme, trying to define the lower-level conditions mechanically, cannot explain these achievements.

> "[...] *these achievements are personal facts which are dissolved by any attempt to specify them in impersonal (or not sufficiently personal) terms. The unspecifiability of such achievements can now be seen to represent a generalization of the theorem that the elements of a commitment cannot be defined in non-committal terms."* (Polanyi 1958, 379)

If we still try this, ultimately, we will question our own personal commitments, beliefs, and knowledge, that emergent reality which we ourselves are also the integral part of (see ch. 13). "Then man dominates a world in which he himself does not exist." (Polanyi 1958, 380)

Polanyi has investigated the gradual levels of evolutionary emergence, starting with the level of reality; the second is the level of growth; the third is the level of (fixed) functionality; the fourth is the level of knowing (learning and problem-solving), and the last is the superior knowledge of humans (based on articulated culture).

> *"The whole ontology of commitment and of a free society dedicated to the cultivation of thought by responsible commitments of its members can in fact be built up, in this manner, as a generalization of biology followed by reflection on this generalized biology."* (Polanyi 1958, 380)

Polanyi's main goal was to reject the empty objectivist scheme and try to free the universal ideals and beliefs as well as the heuristic achievements and commitments of free societies from the captivity of the objectivist critique.

Chapter 13

The rise of man

Goals of the chapter

1. Polanyi continues and deepens the synthesis of epistemology and biology to "re-equip man with the faculties which centuries of critical thought have taught them to distrust." (Polanyi 1958, 281) Unfortunately, we do not believe in our tacit bodily and intellectual skills anymore, despite the fact that they endow us with the ability to know the world.

2. He states that the world is layered with hierarchical levels of reality built upon each other, which is the consequence of evolution: higher-level ones emerge from lower-level ones, and ultimately, every aspect of reality—including both the natural world and the cultural world of man—has arisen from the inanimate material world. Polanyi describes this phenomenon through the concept of emergence. Although he used it in previous chapters, he properly introduces the concept here.

3. Polanyi explains the concept of emergent evolution. However, his goal is not to present a detailed theory concerning the natures of things but only to support his theory of personal knowledge against the ruling materialist and objectivist approaches.

4. Perhaps the most important feature of his vision of evolution is that it does not comply with the neo-Darwinian theory. Polanyi clearly claims that the neo-Darwinian theory as a theory of evolution is false. Still, he fully accepts Darwin's theory of natural selection.

5. He differentiates between the main types of emergence (conceptual and existential) and identifies the major evolutionary levels (life and culture).

6. He expands the concept of morphogenetic fields introduced in the previous chapter to more comprehensive emergent processes of evolution.

7. He explains the rise of man—"anthropogenesis"—from primordial inanimate matter; however, because of the size of the topic, he only explains his concept broadly.

In this chapter, Polanyi first shows that the great transitions in evolution should be understood as the achievements of living beings; his example is the emergence of human consciousness (subch. 2). Then, he defines the process of emergence (subch. 3–4). And finally, he shows the process of emergent evolution (subch. 5–7).

***

> *"We must face the fact that life has actually arisen from inanimate matter, and that human beings [...] have evolved from tiny creatures resembling the parental zygote in which each of us had his individual origin. I shall meet this situation by re-establishing within the logic of achievement, the conception of emergence first postulated by Lloyd Morgan and Samuel Alexander."* (Polanyi 1958, 382)

BOX 23: British emergentism

Emergentism as an independent ontological tradition unifying the advantages of dualism and materialist monism was established by Samuel Alexander in his *Space, Time, and Deity* (1920). Nonetheless, the first emergentist is often considered to be John Stuart Mill, and the term comes from George Henry Lewes's book *The Problems of Life and Mind* (1875). However, these authors were not emergentist at all; they just used a few terms—e.g., Mill's "homeopathic laws" and "heteropathic laws" (*System of Logic* 1843)—which can be regarded as antecedents of later emergentist concepts.

Nevertheless, emergentism did not become a real independent philosophical tradition even after Alexander because, firstly, it was swept away by the rising power of positivism and materialism. Secondly, none of Alexander's most important followers could acknowledge his starting point: reality in its deepest fundaments is *space* and *time*; therefore, even matter itself is an emergent aspect of reality. Both Lloyd Morgan and C. D. Broad started emergence by the *chemical* level based on the fundamental material level, while Alexander himself did not even claim that the chemical level is emergent (Alexander 1920 II.: 61), he instead focused on life and mind, based on the fundamental level of space and time. The main emergent levels are the following, according to Alexander: *matter, life,* and *mind.* So in a sense, space and time are reality itself, the dynamic nature of reality comes from the nature of time, and even space and time are in an emergent relationship with each other.

Lloyd Morgan, in his *Emergent Evolution* (1923), took an approach that can easily be understood as a kind of dualism. Morgan's emergent levels are: *chemical, life, mind,* but over the process of emergence, there is *God* himself. So while Alexander uses the term "deity" to make it clear that he is speaking about a finite, higher-level comprehensive reality emerging in the social/moral spaces of human persons/minds and not about the traditional dualistic and infinite concept of God (Alexander 1920 II.: 347, 361-362), in Morgan's concept, there is God, and his "activity" and involvement from above plays a crucial role in "emergence." Morgan used these particular dualistic concepts to simplify and make Alexander's theory easier to understand, but eventually, it led to a dead end.

Broad's main emergent levels are the following: *chemical, life,* and *mind.* Broad, breaking with Alexander, regarded the unpredictability of certain higher-level phenomena as the reason for emergence, which is an epistemological, strictly scientific category. Consequently, in Broad's theory, emergence is simply contrary to reduction. As a matter of fact, he rejected Alexander's concept because Alexander claimed that mind and other emergent phenomena, which are all the manifestation of time in different levels of spaces, in a (non-materialist, diachronic) sense, could be reduced. (Broad 1925: 639) Broad's main example of emergence is the chemical level, and according to his intentions, contrary to Morgan's "*emergent dualism*" and Alexander's "*emergent neutralism,*" he tried to construct a scientifically more well-established concept of emergence with his concept of "*emergent materialism.*" Broad, contrary to Morgan, put aside the "metaphysical" Alexander and, due to the rising power of logical positivism, used highly analytical methods to reconcile ice with fire. That is, the anti-metaphysical positivist spirit with the original emergentist intention to create a genuinely new ontology, which did not solve the problems, just laid down the foundations of the "emergent materialist" framework in which we tend to speak about emergence still today.

Polanyi, in the fifties, ignores this so-called non-reductive materialist framework and reaches back to the original ontological concept of emergence, reinterpreting it *within the logic of achievement.* He also ignores Alexander's theory of emergent space and time and regards matter as the fundamental level of existence. Accordingly, in Polanyi's view, there are two main emergent levels: *life* and *human culture.*

Evolution as an achievement

Polanyi does not establish a detailed theory of evolution and emergence as, for example, Alexander does, but still clearly commits himself to the ontology of emergence and the process of evolution. The logic of achievement is the main concept by which Polanyi surpasses both the original Alexandrian concept of emergence and the Darwinian theory of natural selection (see ch. 11 and 12). Accordingly:

> *"Lower levels do not lack a bearing on higher levels; they define the conditions of their success and account for their failures, but they cannot account for their success, for they cannot even define it."* (Polanyi 1958, 382)

Polanyi draws attention to the weaknesses of the neo-Darwinian theory (which has been already discussed in ch. 3), emphasizing that the neo-Darwinian theory is not able to account for the evolution of living beings; that is, the emergence and cumulative accumulation of adaptive novelties. The main reason for this deficiency is that the neo-Darwinian approach tries to define living beings and their evolution by objective physical and chemical concepts that are not suited for this task. Natural selection is the consequence of random variants by mutations and random environmental resources, so in itself, due to the nature of lower-level random processes (see ch. 3), it could even lead to degeneration (see in detail in the Box below). Accordingly, neo-Darwinians use the term "evolution" as a deceptive substitution (see ch. 1) because in their theory, "evolution" means only change. Polanyi accepts the Darwinian theory of natural selection that mutations can create advantageous adaptations in terms of reproduction. However, he regards these changes as only one kind of less important events ("conditions") of evolution by which the comprehensive cumulative accumulation of new skills cannot be explained.

BOX 24: The structure of natural selection

According to the neo-Darwinian theory, there is only one fundamental mechanism of evolution, the Darwinian notion of *natural selection.* This is the driving force behind the environmental adaptation of living beings. On the one hand, during inheritance, *new variants* are created by mutation, and, at the same time, there are—generally but not necessarily—*limited environmental resources.* When they exist, these two factors generate the selection race among the different variants: in general, those variants that can utilize the limited environmental resources more successfully will survive and reproduce. With this mechanism, Darwin successfully explained how new life forms are created from old ones. However, he did not explain, on the one hand, how life was originated, since natural selection with its first factor *a priori* presupposes the existence of life, and, on the other hand, why living beings come to be more and more developed. The selective pressure on living beings is determined by the environmental resources that, in turn, ultimately depend on random material processes such as the thermal radiation of the sun, the inclination of the earth's axis, or the drifting of the continents, etc. Although these variables can be described very objectively and precisely, they do not and cannot configure any developmental processes since they are entirely random (see ch. 3). The other factor, mutation, is also random, and in most cases, it does not lead to more developed variants but, in fact, to especially weak, or even unviable, individuals. This means that, over time, an adaptation process governed by natural selection can be reversed. A species might, for example, lose its ability to see or to fly, abilities that most biologists regard as the greatest achievements of evolution. Darwin himself was precisely aware of these problems and did not claim that he had solved the puzzle of the origin of life or even used the term "evolution" for his theory of natural selection.

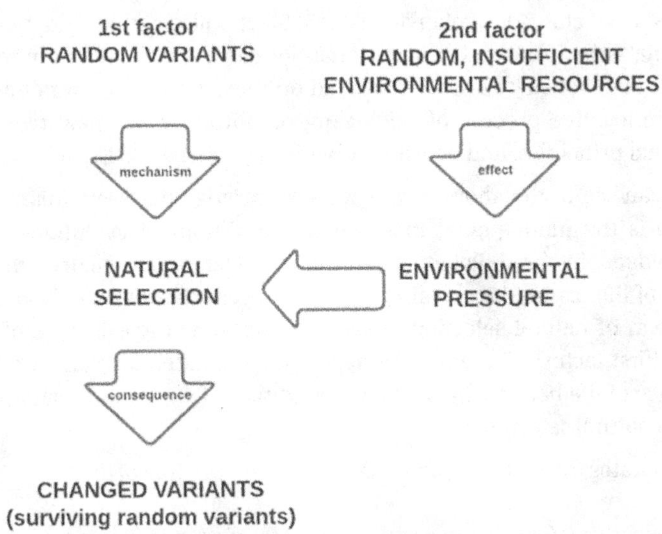

Figure 13.1: *The logical structure of the Darwinian notion of natural selection*

Polanyi translates the problem to ontogeny: there is no explanation for why the random changes of genes by natural selection lead to the more complex and deep structures of living beings during subsequent generations. Every cell of a multicellular organism contains the same genetics. What are the principles that govern why one cell becomes an optic nerve and the other a part of the liver? We know the mechanisms very well, the investigations of which neo-Darwinism is successful beyond dispute, but we do not understand the comprehensive principles at all. Similarly, after a century and a half, we know the mechanism of natural selection perfectly, but not the comprehensive principle, according to which living beings become more and more developed. (Polanyi 1958, 383)

Polanyi's answer to this controversy is that as long as we insist on defining the phenomena of life and evolution by strict objective, physical and chemical terms, we will never have full explanations for these phenomena. We either reject these phenomena, claiming that there is no evolution and there is no life because these phenomena are illusory without any existential meaning—as e.g., the W-scheme by Cassiopeia (see ch. 3)—or we cover up the tension between our theory and reality by deceptive substitutions. Since, in practice, the position that evolution and living beings, including ourselves, do not exist cannot be held, our choice will always be the second option. (Polanyi 1958, 383)

In contrast to the neo-Darwinians, who think according to their objectivist and materialist convictions, Polanyi regards living beings as "instances of morphological types and of operational principles subordinated to a center of individuality," and since living beings act according to the logic of achievement, they cannot be defined by lower-level physical and chemical concepts (see ch. 12). (Polanyi 1958, 383) It follows that "no types, no operational principles, and no individualities can ever be defined in terms of physics and chemistry". Therefore, based on these terms, we cannot account for the innovative process of phylogeny or ontogeny, for new types, new operational principles, and new individualities. (Polanyi 1958, 383)

As we can see in the above box, if we acknowledge that Darwinian natural selection is the fundamental mechanism (condition) of evolution—as it is acknowledged by both Polanyi and the neo-Darwinian theory—then the creation of life cannot be explained by the process of evolution, because the mechanism of natural selection already presupposes the existence of life as variants (first factor). Therefore, we need an independent explanation for the emergence of life because life is the precondition of any evolutionary process based on natural selection.

Polanyi states the following about the ordering principle of life:

"It is clear that for such an event [the first formation of life] to take place two things must be assured: (1) Living beings must be possible, i.e. there

> *must exist rational principles, the operation of which can sustain their carriers indefinitely; and (2) favorable conditions must arise for initiating these operations and sustaining them. In this sense I shall acknowledge that the ordering principle which originated life is the potentiality of a stable open system; while the inanimate matter on which life feeds is merely a condition which sustains life, and the accidental configuration of matter from which life had started had merely released the operations of life."* (Polanyi 1958, 383-384)

So the formation of life from inanimate matter has two preconditions:

First, living beings *have to be possible.* This means that if living beings come into existence, then, besides the necessary lower-level material conditions, there will be higher-level rational principles (rules of rightness), the operation of which can sustain these living beings. Living beings are not material; their (tacit) knowledge works in accordance with these higher-level principles.

Second, *favorable material conditions have to be formed.* This means that if this happens, these favorable lower-level conditions will initiate and then sustain the operation of those higher-level rational principles, which, in turn, sustain the living beings that have come into existence during the process.

The first point might appear to be a tautology, but it is not because, as we will see, it has surprising ontological consequences according to the logic of emergence. So Polanyi regards the *potentiality* of a stable open system as the ordering principle that originated life.

Life itself *is a stable open system.* It has (1) emergent *knowledge* (skills) that works due to higher-level rational principles and (2) fundamental *material conditions.* Before the first formation of life, there were no living beings, only the primordial material conditions and their random fluctuations in space-time. These material conditions and their random fluctuations, however, in themselves cannot initiate the formation of life. As a matter of fact, in themselves, these material conditions are entirely deterministic because they can be said to be random only in relation to other levels, as is the case with mutations in the process of evolution (see ch. 3). In an entirely material world, as materialism conceives it from a Laplacian point of view, *there is no randomness and no potentiality.* In such a world, there are merely the mechanical processes of the fundamental material substance and, consequently, there is no possibility for the emergence of life: living beings are just not possible.

It follows that the first precondition of the emergence of life—i.e., living beings have to be possible—is not a tautology. It is about the true hidden nature of reality. Before the emergence of life in a material world, according to the real emergent nature of reality, there was already *possible* that certain, specific—from the higher level, random—material complexes—that is, certain favorable material conditions—could be formed over time, which would initiate the operation of higher-level rational principles and the formation of life.

So the origin of life is a *potentiality*: life is the consequence of this potentiality. The origin of life is not randomness or a logical necessity. Its cause is not a material process or a vital substance. Life is a possible new, emergent aspect of reality, which, according to the logic of emergence, arises from the randomness of fundamental material conditions in an *unpredictable* and *indefinable* way. In the ontological sense, before the emergence of life, actually, there is merely primordial inanimate matter; but potentially there is life itself because matter is only one aspect of reality, which at a certain point evoked the realization of this potentiality.

According to the second precondition of the emergence of life, favorable lower-level material conditions also have to be formed. This means that *before* the emergence of life, the potential ordering principle of life is nothing more in space-time than a *favorable order of material conditions* which, as a possibility-condition, initiated the emergence of life. Although the ordering principle of life is not an independent ontological existent *before* the emergence of life, just a potentiality in a specific order of matter, it *still cannot be identified* with the fundamental material substance. In this epistemological sense, no order of matter can be identified with matter itself. This means that any favorable order of material conditions initiating the emergence of life is an *epistemologically emergent* phenomenon of the higher level, a new space—we are literally speaking about the specific space and material complexes of the ancient earth—created by this favorable order of material conditions in which the emergence of life can take place.

BOX 25: Epistemological and ontological emergence

In Samuel Alexander's theory of emergence, it becomes clear that different entities might be emergent in different ways. Following Polanyi's example, a crystal is an entity that cannot be accounted for with Laplacian knowledge only, as tacit knowledge is necessary for explaining it, yet it is not emergent in an ontological sense because it is not governed by operational principles and not has to be explained by the logic of achievement. This kind of entity is called epistemologically emergent. The concepts of physics and chemistry cannot describe them, but they are fully material in an ontological sense. As we will see soon, Polanyi, in these cases, talks about *conceptual* emergence. A machine or a living being, on the other hand, has an overarching structure that is also indescribable by Laplacian objective knowledge; moreover, it is also subject to operational principles, and therefore it can only be explained by the logic of achievement. As these entities cannot be said to be material, they are ontologically emergent. Polanyi calls this *existential* emergence.

Polanyi applies the same train of thought to evolution, and he states:

> *"And evolution, like life itself, will then be said to have been originated by the action of an ordering principle, an action released by random fluctuations and sustained by fortunate environmental conditions."* (Polanyi 1958, 394)

Consequently, the ordering principle of evolution is nothing other than *the evolutionary system itself*, that is, the earth. It is a stable open system in the epistemological sense, an emergent comprehensive order of the fundamental material substance, which, due to the logic of emergence, initiates the evolutionary emergence of life; that is, initiates the realization of the possibility of evolutionary emergence in living beings (see below).

The most obvious facts of evolution, according to Polanyi, are the increase in intelligence and the appearance of thinking. By emphasizing this, Polanyi progresses on to the emergent leaps on the higher levels of living beings. He states that he cannot be skeptical of his own consciousness and cannot question the consciousness of others.

> *"Some say that we merely speak in two different languages when referring to thoughts on the one hand and to neural processes on the other. But we speak in two languages because we are talking of two different things."* (Polanyi 1958, 389)

We may talk about the material preconditions of consciousness (on a lower level) or the ontologically emergent functioning of consciousness. We could only say that these are the same if we could reduce the latter to the former. However, this has never been done. It would require the reduction of personal facts to objective facts, and therefore this reduction is impossible—if we accept the reality of our consciousness. Thus, attempts at this reduction will result in mere acts of deceptive substitutions as we (explicitly) dismiss our own (tacit) beliefs, commitments, and experiences. However, we can accept the existence of these—and avoid deceptive substitutions—if we acknowledge that living beings are examples of morphological types and centers of self-interest governed by operational principles. (Polanyi 1958, 385) In other words, by acknowledging personal facts and knowledge, which, however, would not be facts or knowledge in the language of physics and chemistry. Also, we can only claim based on personal knowledge that certain strands of evolution are more successful than others. As a corollary, discussion of evolution is not possible without the acknowledgment of personal knowledge. (Polanyi 1958, 385)

The processes of evolution cannot be observed in short periods of time. Moreover, a lot of random changes are co-occurring with these important processes and thereby mask them. Nonetheless, these random changes are not useless to evolution because they provide the conditions for evolutionary emergence. These random changes can be accounted for by lower-level criteria and without referencing overarching evolutionary trends. Genetics does exactly that, with considerable success in the investigation of random changes common in evolution, while wrongly dismissing the rare and truly significant changes that can be called the achievements of evolution. These are irreducible to lower-level conditions.

> *"But I deny that any entirely accidental advantages can ever add up to the evolution of a new set of operational principles, as it is not in their nature to do so."* (Polanyi 1958, 385)

The application of this to natural selection means the following:

> *"[...] long-range evolutionary progress—like the rise of human consciousness—cannot be determined merely by their adaptive advantage, since these advantages can form part of such progress only in so far as they prove adaptive in a peculiar way, namely on the lines of a continuous ascending evolutionary achievement."* (Polanyi 1958, 385)

The theory of natural selection sees *every* permanent change as adaptation. This becomes a problem since it means that we will miss the discovery that the evolutionary process is also adaptive in the sense of *overarching evolutionary achievements.* Polanyi's example for less overarching change is the color patterns of animals that help camouflage. This is not determined by operational principles and can truly be explained by random change and survival of the fittest. (Polanyi 1958, 401) However, the same cannot be said, for instance, of the emergence of human consciousness, the explication of which requires the sustained effect of the operational principles of evolution.

The mechanisms of mutation and natural selection are only necessary preconditions that need to be present for the operational principles of emergent evolution to be enabled. (Polanyi 1958, 385)

> *"Darwinism has diverted attention for a century from the descent of man by investigating the conditions of evolution and overlooking its action. Evolution can be understood only as a feat of emergence."* (Polanyi 1958, 390)

The emergence of human consciousness

If we want to understand emergent evolution based on the logic of achievement, we cannot dismiss our tacit powers and our personal beliefs—in other words, we cannot make progress without commitments (see in ch. 12). To know that man is the pinnacle of evolution or that it is the most advanced species requires more than (non-Polanyian) natural science. This can only be stated on the grounds of ultrabiology. If we dismiss all of the above and attempt an explanation based on physics and chemistry alone, we are doomed to fail. The process of evolution then becomes a series of random mutations, while the term "evolution" itself becomes a deceptive substitution. As such, it is inevitable to question the existence of our own achievements, including knowing. (Polanyi 1958, 385)

Polanyi calls the path from the first primitive prokaryotes to a human being alive today *anthropogenesis.* This is a clear chain of events with identifiable ancestors. In a direct lineage like this, the investigation of all the involved branches and populations—and there are millions of them—is just a distraction from the true evolutionary achievement on which Polanyi puts all the emphasis.

> *"The bodies of the successive generations of metazoa comprised in one man's ancestry appear as mere carriers of a continuously surviving germ plasm. [...] We may thus regard an anthropogenesis in its entirety as a continuous proliferation of germ plasm, from unicellular origins down to the germ plasm of the human couple of whom the man in question is born."* (Polanyi 1958, 386)

Polanyi defines the following steps of the working of operational principles regarding the living order of the ecosystem. The principles are not *causing* these processes; rather, as we will see later, living beings perform the process of evolution by their own adaptive achievements, by enabling evermore operational principles of living beings. (The Roman numbers below refer to the two big phases of evolution, while the Arabic numbers refer to the important steps within those phases.)

- I.1 The appearance of the *first primitive center of self-interest.* These living beings are mere systems of compartments, and their biggest achievements were to preserve and pass on their internal structure. (Polanyi 1958, 387)
- I.2 *Unicellular organizations.* These were able to move on their own and to perform various activities oriented at self-preservation. Polanyi believes that these were the first undeveloped instances of personal lives.
- I.3 *Multicellular organizations.* Polanyi emphasizes the appearance of sex in reproduction, which, together with the possibility of natural death, increases the personal character of living beings. It is worth mentioning that he imagines that multicellular organizations came about as a result of the clustering of unicellular organizations.
- I.4 The emergence of the *neural system* and *consciousness,* which leads to a deepening of personal characteristics. The evermore advanced neural systems result in more and more consequential decision situations creating subject-object polarity. (Polanyi 1958, 388)
- II.5 The emergence of *intellectual/cultural life.* Following Teilhard de Chardin, Polanyi calls this *noosphere.* This means the appearance of language that brings about a more permanent and articulated frame of thinking, or in other words, a new "material of life."

> *"[...] noogenesis created a new fabric of life not centered on individuals and transcending the natural death of individuals. When man participates in this life his body ceases to be merely an instrument of self-indulgence and becomes a condition of his calling. The inarticulate mental capacities developed in our body by the process of evolution become then the tacit coefficients of articulate thought. By the forming and assimilation of an articulate framework these tacit powers kindle a multitude of new intellectual passions."* (Polanyi 1958, 389)

While Polanyi sees all of the above as important, he emphasizes two leaps: the emergence of life (I.1); and the emergence of culture (II.5). The former marks the beginning of the biological phase, and the latter the cultural phase of evolution.

> *"While the first rise of living individuals overcame the meaninglessness of the universe by establishing in it centers of subjective interests, the rise of human thought in its turn overcame these subjective interests by its universal intent."* (Polanyi 1958, 389)

Unfortunately, Polanyi does not go into details regarding the second phase. Based on Polanyi's focus on articulation, we could infer, however, that the further steps correspond with the further development of linguistic tools. Therefore, we might think of literacy (II.6), then book printing (II.7), and, finally, electronic communication (II.8) as steps that all led to new symbols, formalisms, asynchronous communication channels, and recording devices.

Emergence

The concept of emergence originated with the so-called British emergentists (see BOX 23) who were thinking about, in addition to the fundamental level, higher levels of existence, structures, and order that arise from a substantially material level. Consequently, some natural phenomena exhibit features or a nature that cannot be identified with the sum of their parts.

The British emergentists were not thinking along the lines of the logic of achievement or personal knowledge. Usually, they were thinking about objective emergent laws creating emergent entities. Also, emergence is not only used in a strong, ontological sense but also in an epistemic sense (see BOX 25). In the latter case, we do not talk about emergent entities, only emergent features of material entities.

First, Polanyi attempts to provide a definition of emergence that includes both approaches but in a different sense. In this very general sense, he claims that emergence means that a phenomenon is *undetermined by its details.* Then, emergence, in this sense, means that the whole is in a *random relation* with its details. Therefore, the precondition of emergence is randomness between the whole and the parts; and the appearance of randomness in the world is to be seen as the most basic example of emergence. (Polanyi 1958, 390)

> *"I call this a case of emergence, for we can know the randomness of a system, yet cannot know it in terms of a more detailed knowledge of the system. Our knowledge of this emergent quality, randomness, is in fact destroyed by observing the particulars which determine the system below the emergent level."* (Polanyi 1958, 391)

Polanyi explains this using the example of heat. This choice is fortunate in terms of providing contrast as heat is also a usual example of randomness and reduction in physics. This shows that reduction is not wholly incompatible with emergence.

> *"The gas can be said to have a definite temperature and a definite pressure only if we assume that its molecules are in random motion; an assumption which is incompatible with our knowing the configuration of molecular motions in the gas."* (Polanyi 1958, 391)

> *"If we knew exactly the position and velocity of each molecule (within the limits of wave mechanics) we could only predict the behaviour of the molecules, but not the comprehensive features defined by randomness."* (Polanyi 1958, 391)

Naturally, gas particles do not move around randomly. Theoretically, it is possible to give a full account of each and every particle moving in a system. But this knowledge, in turn, cannot give an account of comprehensive properties such as pressure and temperature. There is no randomness at all allowed; the movement of each particle is determined. Knowledge of it all would be an example of Laplacian knowledge. (Polanyi 1958, 394) We can realize that our relationship to the movements of gas particles is similar to that of the "random" pebbles forming the fake "Wales" scheme from ch. 3.

The movements of gas particles are only random in the context of the whole, comprehensive system of the gas—and only the gas as a whole has pressure and temperature. The knowledge about this is a personal impression that has nothing to do with the Laplacian knowledge of the actual particle movements, which in this lower-level context is perfectly deterministic to the laws of physics.

However, this higher-level knowledge might be reduced to lower-level knowledge if the overall feature can be fully accounted for with the said movements. But if the overall feature cannot be accounted for with such movements (such as in the case of the real lettering of "Wales," as it has been arranged by a person), then this reduction is impossible. Gas is not such a case because temperature can be understood as the average energy of the moving particles. Reduction destroys the existential meaning of higher-level knowledge. So this case is analogous with case 3 of the Wales example from ch. 3, where the Wales text has no meaning; it is only a random illusion.

To sum up, we understand temperature to be a higher-level consequence of the deterministic movement of the gas particles only, while, from the point of view of the gas as a whole, these movements are random. (Polanyi 1958, 392)

It is a mistake to think that there is no difference between the lower-level Laplacian knowledge and higher-level personal knowledge. In the case of gas, we have both kinds of knowledge; moreover, the successful reduction seems to remove the difference altogether. In a Polanyian example, casting a dice is completely deterministic in itself, yet we cannot predict its average outcome. That is an element of personal knowledge, i.e., knowing that the individual outcomes are random in relation to their average of 3.5. And therefore, by knowing every detail of each individual act of dice casting, the average cannot be calculated. That can only be done in the act of personal appraisal. After a series of attempts, we might calculate an exact average—e.g., 3.745523642442—but that will not be the same—or will be the same only accidentally—as the outcome of the personal appraisal, which is, of course, always 3.5. If we dealt with probability in this manner, we would be forced to accept this figure or make more attempts but never have an indication to settle. However, we do not deal with probability in that manner. We look at the six-sided symmetry of the dice and the relation of this structure with the physical process of throwing the dice, and we conclude that the expected average is 3.5. These two types of knowledge are fundamentally different. (Polanyi 1958, 391)

Material systems exhibit many comprehensive, epistemologically emergent properties—rightness or order—that cannot be grasped with Laplacian knowledge and are incomprehensive on that level. However, many of these features would be destroyed by entropy if not for the higher-level operational or ordering principles that, for example, in the case of life and evolution, keep the degrading effects of entropy in check. (Polanyi 1958, 391)

> *"The emergent form of existence is identified by our comprehensive judgment of it, which judgment accredits thereby, indirectly, a correlated context of properties, and of problems and manipulations, all of which presuppose the emergent form of existence and serve to elaborate its reality."* (Polanyi 1958, 392)

There are two modes of existence, then:

1. *Substantive material,* the processes of which can be described by Laplacian knowledge;
2. *Emergent systems*—that is, systems of rightness that are appraised by personal knowledge.

Matter always has an existential meaning; it is an existing entity in an ontological sense: it is a substance. Emergent systems, however, do not always exist in the ontological sense, such in the case of heat, and in this case, they have no existential meaning either—only denotative, describing certain specific comprehensive features of the system such as pressure, for instance.

The question is, then, where is the demarcation between epistemic and ontological emergence? (Polanyi 1958, 392)

> *"The first thing to observe here is that, strictly speaking, it is not the emerged higher form of being, but our knowledge of it, that is unspecifiable in terms of its lower-level particulars."* (Polanyi 1958, 393-394)

Polanyi's typical example is the crystal:

> *"There is, indeed, always a noticeable logical gap between a topography and a pattern derived from it, and to this extent no pattern is specifiable in terms of its topography. Yet since in the case of a crystal, we can easily pass from the pattern to the topography and back again, the conception of such a pattern is in fact not destroyed by a knowledge of its topographic particulars. I would acknowledge, therefore, in this case two distinguishable conceptual levels but not two separate levels of existence."* (Polanyi 1958, 394)

The description of any given higher-level pattern will always differ from the lower-level topography as they use *different* concepts. In the case of the crystal, heat, and others, the higher-level description can always be reduced to the lower-level description. The ordering principle found in crystal formation is found at the material level. This means that, ontologically, the crystal is a material entity while having epistemologically emergent features. (Polanyi 1958, 394)

In connection to this, Polanyi claims that while quantum mechanics can, in theory, completely reduce the entirety of chemical properties, physics can never replace chemistry. Because chemistry deals with questions relevant to a particular personal interest and practical conditions, the more fundamental Laplacian knowledge will think about matter in any conditions and any circumstances. As such, it cannot answer, not even in theory, the questions arising from the personal point of view. Those can only be dealt with by the chemist through his experience, skills, and curiosity.

> *"Therefore, while quantum mechanics can explain in principle all chemical reactions, it cannot replace, even in principle, our knowledge of chemistry. We may acknowledge this as an incipient separation of two forms of existence."* (Polanyi 1958, 394)

Here, the two modes of existence mean the distinction between substantive matter and epistemologically emergent order, which—unlike the case of life—does not yet mean an ontological distinction. However, in terms of knowledge, the two modes require different epistemic toolkits. Chemistry can handle the emergent order, while quantum mechanics only the material parts, and therefore both might be called Laplacian.

As for the question as to when it is justified to talk about a new *existential meaning* or new levels of existence, Polanyi says the following:

> *"No richly endowed new reality can be seen emerging in the inanimate domain. This happens for the first time in the emergence of a living being from inanimate constituents [...] This results in the formation of two levels of existence: an upper level governed by physiology, and a subsidiary, lower level defined by physics and chemistry."* (Polanyi 1958, 394)

In other words, it happens exactly when the center of self-interest is created. Through using its operational principles, this will maintain its own integrity. In the case of living beings, this means ordering principles that are not determined by material conditions but by the logic of achievement. It follows that the *logic of achievement is at the same time the logic of emergence too.* Thanks to the logic of achievement, the relationship between physiology and physics is similar to that between morphology and physics, or logic and psychology (see ch. 12).

Emergent evolution

Polanyi tries to resolve a false dichotomy. One horn of the argument is that, during emergent evolution, a creative agent inserted itself in the process (which would assume the presence of a Creator) or that the whole process was predetermined from the beginning. Any of these would amount to the denial of evolution.

> *"No new creative agent, therefore, need be said to enter an emergent system at consecutive new stages of being. Novel forms of existence take control of the system by a process of maturation."* (Polanyi 1958, 395)

At this point, Polanyi uses a sophisticated pair of concepts to explain what is going on here: *a fronte* (*from the front*) determinism and *a tergo (from behind)* determinism.

> *"The argument will fall into two parts, the first dealing with determinism a fronte by the universal target of a commitment, the other with determinism a tergo by the bodily mechanism of the person entering on a commitment."* (Polanyi 1958, 395)

With the two kinds of determinism, Polanyi lays the foundations of a teleological theory that applies from the first cells to humans without falling into the trap of vitalism. This is important because anyone making claims about the teleological nature of evolution is easily labeled to vitalism, and vitalism is synonymous with an unscientific, unserious approach. This is a result of the rise of neo-Darwinism in the first half of the twentieth century.

A fronte determinism is teleological and follows the logic of achievement. The habitat on Earth is full of order, both in an epistemic and ontological sense. Recognizing, exploiting, and manipulating this order must be essential aims for living beings in order for them to stay alive. This is the way they adapt to their surroundings, described by the logic of achievement. No living being approaches the world with a Laplacian kind of knowledge; they use their personal commitments. This means that they act according to their own personal norms and maxims while accepting emergent phenomena as manifestations of reality.

> *"All the time the creative mind is searching for something believed to be real; which, being real, will—when discovered—be entitled to claim universal validity—something the knowledge of which must indeed passionately insist on its own universal validity."* (Polanyi 1958, 396)

As is always the case with the logic of achievement, living beings sometimes make mistakes in knowing, so their activity can be either success or failure. Moreover, for problem-solving, we need to make various kinds of leaps of belief.

> *"Comprehension is an unformalizable process striving towards an unspecifiable achievement, and is accordingly attributed to the agency of a center seeking satisfaction in the light of its own standards."* (Polanyi 1958, 396)

This means, moreover, that no external factor makes living beings work—no vital force or divine powers. (Polanyi 1958, 396) Living beings are subjects of their own commitments and the commitment of knowing reality. In this sense, this process is teleologically deterministic.

> *"Such is the logic of self-compulsion with universal intent. Action and submission are totally blended in a heuristic communion with reality; determinism and spontaneity mutually require each other when embodied in the universal and the personal poles of commitment."* (Polanyi 1958, 396)

There is also *a tergo* determinism which is not teleological. This means the effect of those material conditions that initiate and maintain the higher-level operational principles. These conditions need to be there but cannot fully determine the actions of living beings.

> *"Thus the personal pole of commitment retains its autonomy everywhere, exercising its calling within a material milieu which conditions but never fully determines its actions."* (Polanyi 1958, 397)

The emergent order of the habitat of Earth as the universal pole of our commitments cannot be discovered without the person's involvement. Without *a fronte* determinism, emergent evolution might still have an aim, but its spontaneous process would be unexplainable. The knowledge of living beings about reality that has universal intent would cease to be a measure of adaptation and achievement—nothing would be left of the process of evolution but the random consequences of the *a tergo* determinism, which is not in any sense progress. This is why Polanyi claims, "Biology would gain greatly in scope and depth by addressing itself more candidly to the fundamental features of life." (Polanyi 1958, 399) It should not, therefore, look only for material, *a tergo* causes and conditions, as it does currently in its neo-Darwinian tradition.

Polanyi then moves on to place emergent evolution in the framework of a field theory. The goal remains to avoid vitalism; thus, he picks his examples mostly from physics.

> *"Just as mechanical forces are the gradients of a potential energy, so this field of forces would also be the gradient of a potentiality: a gradient arising from the proximity of a possible achievement. Our sense of approaching the unknown solution of a problem, and the urge to pursue it, are manifestly responses to a gradient of potential achievement [...]"* (Polanyi 1958, 398)

The important distinction between a living and a non-living field is that the former is connected to a certain centrum.

> *"Biological fields normally belong to a center to which both the opportunities and the strivings are attributable. Though these strivings are continuous with the conscious strivings of higher animals, they are, of course, in general, neither conscious nor deliberate. By contrast to a field of forces operating on an inanimate system, a field of biological striving stands defined by the fact that we attribute its operations to an active center, and acknowledge these operations as the successes or failures of this center, while basing our awards of success or failure on criteria which are largely unspecifiable."* (Polanyi 1958, 404)

The ordering principle of evolution—as we have already seen—only allows for potential achievements but does not make emergent evolution inevitable, itself being a non-ontological entity, unlike a vital force or a God would be. The order of the habitat of Earth, an epistemically emergent property, is what we call the ordering principle of evolution. This, in turn, means that the potential evolutionary achievements need to be activated heuristically by living beings. Living beings have an inner drive to understand and manipulate emergent phenomena, which drives emergent evolution forward. (Polanyi 1958, 398)

The operators of biological fields comprise three stages of originality:

1. The *originality of resourcefulness*, the solving of a problem;
2. *Ontogenetic maturation*, which repeats the achievements of the phylogenetic maturation but in an individual organism;
3. *Phylogenetic emergence* that produces completely unprecedented operational principles. (Polanyi 1958, 399)

In Polanyi's interpretation, phylogenetic emergence—in other words, emergent evolution itself—is nothing other than the maturation of the possibilities of ontogenesis. (Polanyi 1958, 399) Ontogenesis is the actual process by which an emergent organism develops from the initial germ in given conditions that might be limiting. Ontogenetic maturation results in a certain capability of the individual organization for originality.

The bottom line is that anthropogenesis is about the story of the germ plasm as "the evolutionary process takes place in the germ plasm, but it manifests itself in the novel organism which the germ plasm potentially embodies." (Polanyi 1958, 399) However, the development of the germ plasm is not limited in the same way as the development of the organization, "[...] the possibility of unprecedented achievements guides the maturation of the germ plasm to ever higher evolutionary stages." (Polanyi 1958, 400)

The development of the germ plasm is enabled by the ordering principle of evolution—in other words, Earth's epistemically emergent order—that presents the organisms with certain problems and tasks that motivate them (in *a fronte* sense) to live up to those tasks through achievements in individual heuristic acts as well as ontogenesis.

Conclusion

In the last chapter, Polanyi presents emergent evolution, which is the precondition of personal knowledge. To be more accurate, the possibility of personal knowledge was created by evolution and therefore accepting emergent evolution means accepting the personal nature of ourselves and our knowledge. As a corollary, evolution cannot be discussed with impersonal, supposedly objective concepts.

> *"An anthropogenesis induced by random mutations would then be essentially on a par with the ordering of a set of biased dice under the impact of Brownian motion at low temperatures."* (Polanyi 1958, 402)

In other words, according to the neo-Darwinian view, the impossible is possible. That is, it is possible to cast sixes all the time with unloaded dice (see ch. 3): the organisms realizing the higher-level forms of order—the case of the all-sixes in the dice analogy—can appear without higher-level ordering principles. This mode of explanation of evolution leads to a reduction of the achievements of evolution—the organisms and also the very person who proposes this view—to nothing, including the fact that we hide this with deceptive substitutions.

In Polanyi's view, the spine of the achievements of evolution is the emergent processes of animal knowing.

> "[...] that knowing belongs to the class of achievements that are comprised by all forms of living, *simply because every manifestation of life is a technical achievement, and is therefore—like the practice of technology—an applied knowledge of nature.*" (Polanyi 1958, 403-404)

> *"At all levels of life it is these centers which take the risks of living and believing. And it is still such centers which, at the highest stage of development, actuate those men who seek the truth and declare it to all comers—at all costs."* (Polanyi 1958, 404)

Living beings have the capability and the drive to discover the ordered nature of the habitat of the planet, which presents them with problems that can be overcome by achievements. We are talking here about the personal adaptive skills that bring evolution to its next steps: towards unforeseen achievements, commitments, beliefs, and knowledge. Mutations and the impersonal, lower-level events of natural selection cannot account for those; they merely provide the preconditions of the rise of man.

Bibliography

Alexander, Samuel. *Space, Time, and Deity.* London: MacMillan and Co., 1920.

Broad, C. D. *The Mind and its Place in Nature.* New York: Routledge, 1925.

Kepler, Johannes. *The Harmony of the World.* American Philosophical Society. 1997.

Laplace, Pierre-Simon. *A Philosophical Essay on Probabilities.* New York: J. Wiley and Sons, 1902.

Lewes, George Henry. *Problems of Life and Mind. First Series. The Foundations of a Creed Vol. II.* Boston: James S. Osgood, 1975.

Mill, John Stuart. *System of Logic.* London, Longmans, Green, Reader, and Dyer. 1843.

Morgan, C. Lloyd. *Emergent Evolution.* London: Williams and Norgate, 1923.

Polanyi, Michael. "Life's Irreducible Structure." In *Michael Polanyi: Knowing and Being: Essays,* ed. Marjorie Grene, 225–39. New Brunswick: Transaction, 1969.

———. "Logic and Psychology." *American Psychologist* 23 (1968) 27–43.

———. *Personal Knowledge.* Chicago: University of Chicago Press and London: Routledge and Kegan Paul, 1958.

———. *The Study of Man.* London: Routledge and Kegan Paul, 1959.

Bibliography: Polanyi's books and important articles

Michael Polanyi. *USSR Economics.* Manchester: Manchester University Press, 1935.

———. *The Contempt of Freedom.* London: Watts and Company, 1940.

———. *Full Employment and Free Trade.* Cambridge: Cambridge University Press, 1945.

———. *Science, Faith and Society.* Chicago: University of Chicago Press, 1946.

———. *The Logic of Liberty.* Chicago: University of Chicago Press, 1951.

———. *Personal Knowledge.* Chicago: University of Chicago Press, 1958.

———. *Beyond Nihilism.* London: Cambridge University Press, 1960.

———. *The Study of Man.* Chicago: University of Chicago Press, 1964.

———. *The Tacit Dimension.* New York: Doubleday and Company, 1966.

———. "Science in U.S.S.R." *The New Statesman and Nation* 19 (1940) 174.

———. "Cultural Significance of Science." *Nature* 147 (1941) 119.

———. "The Growth of Thought in Society." *Economica* 8 (1941) 428–456.

———. "Research and Planning." *Nature* 152 (1943) 217–218.

———. "Science—Its Reality and Freedom." *The Nineteenth Century and After* 135 (1943) 78–83.

———. "The Autonomy of Science." *Memoirs and Proceedings of the Manchester Literary and Philosophical Society* 85 (1943) 19–38.

———. "Patent Reform." *The Review of Economic Studies* 11 (1944) 61–76.

———. "Science – Its Reality and Freedom." *The Nineteenth Century* 135 (1944) 78–83.

———. "Science and the Modem Crisis." *Memoirs and Proceedings of the Manchester Literary and Philosophical Society* 86 (1945) 7–16.

———. "Reform of the Patent Law in Britain." *Nature* 156 (1945) 54.

———. "The Autonomy of Science." *The Scientific Monthly* 60 (1945) 141–150.

———. "The Foundations of Freedom in Science." *The Nineteenth Century* 141 (1947) 163–167.

———. "Science: Observation and Belief." *Humanitas* 1 (1947) 10–15.

———. "The Logic of Liberty: Perils of Inconsistency." *Measure* 1 (1950) 348–62.

———. "The Hypothesis of Cybernetics." *The British Journal for the Philosophy of Science* 2 (1951) 309–312.

———. "The Stability of Beliefs." *The British Journal for the Philosophy of Science* 3 (1952) 217–32.

———. "Science and Faith." *Question* 5 (1952) 15–45.

———. "From Copernicus to Einstein." *Encounter* 5 (1955) 54–63.

———. "The Magic of Marxism." *Encounter* 7 (1956) 5–17.

———. "Problem-solving." *The British Journal for the Philosophy of Science* 8 (1957) 89–103.

———. "Beyond Nihilism." *Encounter* 14 (1960) 34–43.

———. "Faith and Reason." *Journal of Religion* 41 (1961) 237–247.

———. "Knowing and Being." *Mind* 70 (1961) 458–470.

———. "The study of Man." *Quest* 29 (1961) 26–35.

———. "The Unaccountable Element in Science." *Transactions of the Bose Research Institute* 24 (1961) 175–184.

———. "Tacit Knowing: Its Bearing on Some Problems of Philosophy." *Philosophy Today* 6 (1962) 239–262.

———. "The Unaccountable Element in Science." *Philosophy Today* 6 (1962) 171–182.

———. "The Feelings of Machines." *Encounter* 22 (1964) 85–86.

———. "On the Modern Mind." *Encounter* 24 (1965) 12–20.

———. "The Structure of Consciousness." *Brain* 88 (1965) 799–810.

———. "The Creative Imagination." *Chemical and Engineering News* 44 (1966) 85–93.

———. "The Logic of Tacit Inference." *Philosophy* 41 (1966) 1–18.

———. "The Message of the Hungarian Revolution." *American Scholar* 35 (1966) 661–676.

———. "Life Transcending Physics and Chemistry." *Chemical and Engineering News* 45 (1967) 54–66.

———. "The Growth of Science in Society." *Minerva* 5 (1967) 533–45.

———. "Sense-Giving and Sense-Reading." *Philosophy* 42 (1967) 301–325.

———. "Science and Reality." *The British Journal for the Philosophy of Science* 18 (1967) 177–196.

———. "The Creative Imagination." *Tri-Quarterly* 8 (1967) 111–123.

———. "Logic and Psychology." *The American Psychologist* 12 (1968) 27–43.

———. "Life's Irreducible Structure." *Science* 160 (1968) 1308–1312.

———. "On Body and Mind." *The New Scholasticism* 43 (1969) 195–204.

———. "Science and Man." *Proceedings of the Royal Society of Medicine* 63 (1970) 969–979.

———. "Why Did We Destroy Europe?" *Studium Generale* 23 (1970) 909–016.

———. "Genius in Science." *Encounter* 38 (1972) 43–50.

———. "Truth in Myths." *Cross Currents* 25 (1975) 149–162.

Bibliography: Some important books on Polanyi

Allen, Richard. *Thinkers of Our Time: Polanyi.* London: Claridge Press, 1990.

Bíró, Gábor. *The Economic Thought of Michael Polanyi.* London: Rutledge, 2021.

Gelwick, Richard. The Way of Discovery: *An Introduction to the Thought of Michael Polanyi.* New York: OUP, 1977.

Gill, Jerry H. *The Tacit Mode: Michael Polanyi's Postmodern Philosophy.* Albany, NY: SUNY Press, 2000.

Jha, Stefania Ruzsits. *Reconsidering Michael Polanyi's Philosophy.* Pittsburgh: University of Pittsburgh Press, 2002.

Meek, Esther Lightcap. *Contact with Reality: Michael Polanyi's Realism and Why It Matters.* Eugene, OR: Cascade Books, 2017.

Mitchell, Mark T. *Michael Polanyi: The Art of Knowing.* Wilmington, Delaware: ISI Books, 2006.

Nye, Mary Jo. *Michael Polanyi and his Generation.* Chicago: The University of Chicago Press, 2013.

Paksi, Daniel. *Personal Reality. The Emergentist Concept of Science, Evolution, and Culture.* Volume 1&2. Eugene, OR: Pickwick Publications, 2019.

Poirier, Maben W. *A Classified and Partially Annotated Bibliography of Michael Polanyi, the Anglo-Hungarian Philosopher of Science.* Toronto: Canadian Scholars' Press, 2002.

Prosch, Harry. Michael Polanyi: *A Critical Exposition.* Albany, NY: SUNY Press, 1986.

Scott, Drusilla. *Everyman Revived: The Common Sense of Michael Polanyi.* Lewes, Sussex: Book Guild Limited, 1985.

Scott, William Tausig and Martin X. Moleski. *Michael Polanyi: Scientist and Philosopher.* Oxford: Oxford University Press, 2005.

Index

E

F

G

H

I

O

P

Q

R

S

CPSIA information can be obtained
at www.ICGtesting.com
Printed in the USA
LVHW080818140422
716182LV00015B/318/J

9 781648 893131